Cosmology and the Early Universe

Series in Astronomy and Astrophysics

The *Series in Astronomy and Astrophysics* includes books on all aspects of theoretical and experimental astronomy and astrophysics. Books in the series range in level from textbooks and handbooks to more advanced expositions of current research.

Series Editors:
M Birkinshaw, University of Bristol, UK
J Silk, University of Oxford, UK
G Fuller, University of Manchester, UK

Recent books in the series

Dark Sky, Dark Matter
J M Overduin and P S Wesson

Dust in the Galactic Environment, 2nd Edition
D C B Whittet

The Physics of Interstellar Dust
E Krügel

Very High Energy Gamma-Ray Astronomy
T C Weekes

Numerical Methods in Astrophysics: An Introduction
P Bodenheimer, G P Laughlin, M Rózyczka, H W Yorke

An Introduction to the Physics of Interstellar Dust
Endrik Krugel

Astrobiology: An Introduction
Alan Longstaff

Fundamentals of Radio Astronomy: Observational Methods
Jonathan M Marr, Ronald L Snell and Stanley E Kurtz

Stellar Explosions: Hydrodynamics and Nucleosynthesis
Jordi José

Cosmology for Physicists
David Lyth

Cosmology
Nicola Vittorio

Cosmology and the Early Universe
Pasquale Di Bari

For more information about this series, please visit: https://www.crcpress.com/Series-in-Astronomy-and-Astrophysics/book-series/TFSERASTAST

Cosmology and the Early Universe

By
Pasquale Di Bari

CRC Press
Taylor & Francis Group
Boca Raton London New York

CRC Press is an imprint of the
Taylor & Francis Group, an **informa** business

CRC Press
Taylor & Francis Group
6000 Broken Sound Parkway NW, Suite 300
Boca Raton, FL 33487-2742

First issued in paperback 2020

© 2018 by Taylor & Francis Group, LLC
CRC Press is an imprint of Taylor & Francis Group, an Informa business

No claim to original U.S. Government works

Version Date: 20180418

ISBN-13: 978-0-367-57170-2 (pbk)
ISBN-13: 978-1-4987-6170-3 (hbk)

Library of Congress Cataloging-in-Publication Data

Names: Di Bari, Pasquale, author.
Title: Cosmology and the early universe / by Pasquale Di Bari.
Description: Boca Raton : CRC Press, [2018] | Series: Series in astronomy and astrophysics | Includes bibliographical references and index.
Identifiers: LCCN 2018003315 (print) | LCCN 2018005976 (ebook) | ISBN 9781498761727 (eBook - Vitalsource) | ISBN 9781498761710 (eBook - PDF) | ISBN 9781498761734 (eBook - ePub) | ISBN 9781138496903 (eBook - General) | ISBN 9781498761703 (hardback : alk. paper)
Subjects: LCSH: Cosmogony. | Cosmology. | Universe.
Classification: LCC QB981 (ebook) | LCC QB981 .D475 2018 (print) | DDC 523.1--dc23
LC record available at https://lccn.loc.gov/20180033152

Visit the Taylor & Francis Web site at
http://www.taylorandfrancis.com

and the CRC Press Web site at
http://www.crcpress.com

A Franco & Marisa

Contents

Conventions

- Throughout the book, only temperature is in natural units, in a way that the Boltzmann constant $k_B = 1$; for all other quantities we adopt SI units.

- We use $\eta_{\alpha\beta} = \mathrm{diag}(+, -, -, -)$ for the Minkowski metric tensor.

- We indicate pressure and 4-momentum vector with the same symbol p, while the length of 3-momentum vector is indicated with $|\vec{p}|$.

- Greek indexes run from 0 to 3; Latin indexes run from 1 to 3.

- All chapters contain an initial abstract, a final section of exercises and an independent list of references.

Preface

Since the end of the 1990s, cosmology has experienced one of the most impressive advances among all scientific disciplines. This happened mainly because of astonishing progress in the precision and accuracy of astronomical and cosmological observations resulting in a breathtaking sequence of major scientific discoveries.[1] These discoveries made it possible to settle long-standing issues and to pin down a well-defined cosmological model able to describe essentially all cosmological observations: the ΛCDM model.[2] Within the ΛCDM model, the history of the universe can be traced back to a time just a tiny fraction of a second after the singularity, more popularly known as the *Big Bang*. The main aim of this book is to provide the tools needed for an understanding of the basic features of the ΛCDM model and the underlying observations.

There are different motivations today to study cosmology. First of all, it is simply fascinating to understand the fundamental properties and laws governing the universe itself, trying to answer (in quite a few cases successfully) questions that puzzled the best thinkers since ancient times. On more practical grounds, for those whose main interest is astrophysics, a knowledge of the most recent cosmological results and issues, and of the basic features of the ΛCDM model, is very important for a deeper understanding of the astrophysical environments. The history of stars, galaxies and other astrophysical objects is in many cases inextricably linked to cosmological properties. For those who have attended (or plan to attend) a course on general relativity, it will be interesting to discover how Einstein's theory of gravity provides the correct description for a successful cosmological model. Therefore, cosmology can be certainly regarded as one of the most important applications (and triumphs) of general relativity.

However, a description of gravity in terms of general relativity is not sufficient to understand our universe. The properties of the elementary particles and of their interactions are also crucial to build a cosmological model consistent with the observations. Using a theatrical analogy, it is fair to say that if general relativity provides the stage, elementary particles, more precisely the quantum fields, are the actors and both are fundamental to the play. In particular, the standard model provides a very successful description of elementary particles and their fundamental interactions needed (also) for the understanding of the physics of the early universe.

---

[1] As witnessed by two Nobel Prizes in Physics awarded in this period for discoveries in cosmology: in 2006 to G.F. Smoot and J.C. Mather for the discovery of the cosmic microwave background radiation anisotropy and in 2011 to S. Perlmutter, B.S. Schmidt and A.G. Riess for the discovery of the accelerating expansion of the universe through observations of distant supernovae.

[2] This cryptic name will become clear at the end of the first part of the book.

For example, one of its main successes is the explanation of primordial nuclear abundances. Any proposed new model of particle physics, beyond the standard model, must pass many cosmological tests and not spoil the successful cosmological predictions from the standard model. In this way, cosmology places bounds on the parameters of the new tentative model, often severely constraining it and sometimes completely ruling it out.

On the other hand, even more interestingly, there are some important cosmological features of the ΛCDM model that do not seem to be understandable just by combining the standard model with cosmology. The most spectacular example is offered by the *dark matter puzzle*, whose solution seems to point to the existence of a hitherto undiscovered elementary particle. If this cosmological prediction should one day be confirmed by laboratory experiments, it would represent a triumph for cosmology. Thus, cosmological tools provide very important guidance for particle-physics model builders. As a last motivation I would like to emphasise that within cosmology essentially all physical disciplines come into play in a relevant way. Cosmology is a genuine synergic field where one needs to combine different knowledge. Statistical mechanics, fluid dynamics, particle physics, astronomy and astrophysics, general relativity and quantum field theories, join together to give an extremely detailed and rich picture of the history of the universe and of its properties. It can certainly be said that cosmology is a highly unifying field in physics.

The book is divided into two parts: in the first, cosmological observations and the Einstein theory of gravitation, applied to the homogeneous and isotropic universe, are linked together, showing that a flat Lemaitre cosmological model, as the ΛCDM model is, successfully describes the expansion of the universe. In the second part, it is shown how this model, with the proper initial conditions, can be extended to reconstruct crucial periods in the history of the early universe, explaining fundamental cosmological observations that require the occurrence of an early hot stage. In this part the tight connections between the early universe and particle physics will play a central role. This second part can be therefore regarded as an introduction to *particle cosmology*, one of the most fascinating topics in modern physics, where microscopic and cosmological properties need to be beautifully linked together to obtain a consistent unified picture.

It is really a great pleasure to thank all students and colleagues that have contributed many comments, suggestions and discussions to improve the book, in particular: Sacha Belyaev, Kareem Farrag, Jonathan Flynn, Steve King, Patrick Ludl, Tim Morris and Francesco Shankar.

Finally, I would like to conclude with a historical note. It is a hundred years since Einstein proposed his famous cosmological static model, with the introduction of the cosmological constant that marked the birth of modern cosmology. I am certainly pleased to celebrate this important centenary with this book but, at the same time, I hope it will show how cosmology, despite being a hundred years old, is one of the most lively scientific fields with new important discoveries yet to come.

Pasquale Di Bari,
Southampton

I

Cosmology

Historical breakthroughs

The cosmologies of primitive social communities were extremely *local*, tightly based on their local surroundings. They were typically consisting of myths and legends and the universe was usually described in an animistic way. Ancient cosmology is therefore profoundly anthropomorphic. For example, in the Australian aboriginal mythology, the universe, the same space and time as we know it, was created in the fight between two ancestral kinds of Gods with snake appearance. The fight took place around the holy rock Uluru that, therefore, represents a sort of centre of the universe for the Australian Aborigines, certainly a privileged point.

Analogously, almost all ancient civilisations usually attributed to some special places an important role in their life or religious beliefs (Mount Olympus for early Greeks, the Ziggurat Etemenanki for Babylonians and so forth). At the same time astronomical observations were strongly influencing the social life. Stars and astronomical events were regarded as divine signs and a way to determine the future and take decisions on a social or even individual scale. The calendar of the ancient communities could then be used as a medium of social control from the authority.

It is only with Greek philosophers that there was a first attempt to look at the universe as regulated by timeless mathematical and geometrical laws to be studied. Then not by chance the word *cosmology* has a Greek origin that carries a connotation of a *regular behaviour and beauty*. Yet Greek cosmologies had still an anthropocentric character. In the geocentric Aristotelian model, Earth is the centre of the universe. Generally speaking, one can say that the progress in the cosmological description is marked by a gradual *delocalization* process, where Earth and its inhabitants occupy a less and less privileged position in the universe, something often referred to as the *Copernican principle* . This process happened not just because of a progressive separation of the cosmological ideas from religious and philosophical prejudices but also, especially in more modern times, by virtue of an impressive advance in observational tools. These provided a progressively more accurate picture of the structure of the universe.

However, it should be said that the long time needed for the replacement of the geocentric model (Aristotle, Ptolemy) with the heliocentric model (Aristarchus, Copernicus) was not just a matter of philosophical prejudice. It was mainly due to the lack of experimental evidence due to instrumental limitations. In a heliocentric model, one expects to observe, due to Earth's annual motion around the Sun, the so-called stellar annual parallax effect, such that closer stars should be seen moving with respect to stars further in the background. This effect could be observed only in 1838 by the astronomer Friedrich Bessel for the star 61 Cygni [1]. The non-observation of this effect was used as one of the main (absolutely justified!) objections against the heliocentric system, since stars should have distances that at that time were considered absurdly too large. From this point of view, it can be fully appreciated how Copernicus supported the heliocentric model on the basis of merely aesthetic and conceptual reasons: planetary motions could be described in a much simpler way within the heliocentric model.

The problem of measuring distances, and of accepting a universe that is outrageously much bigger than the length scales we experience in everyday life, was one of the crucial obstacles for the advent of a modern cosmological description. This problem persisted even much later, even though at a different level.

Even after the heliocentric model was fully accepted, only in the early twentieth century was it understood, by Harlow Shapley, that our solar system is not placed at or even in the vicinity of the centre of the Galaxy, as claimed by William Herschel in the eighteenth century, but at a distance of two thirds of the galactic radius [2]. This seemed to support the idea that our Galaxy, the Milky Way, much bigger than what was thought until that moment, could encompass the entire universe itself. This was the picture supported in 1920 by Shapley himself, in contrast with the so-called *island universe* theory. In this theory, the observed spiral nebulae are nothing more than copies of our Galaxy of similar size, an idea tracing back to the philosopher Immanuel Kant at the end of the eighteenth century.

This scientific dispute culminated, in 1921, in the *Great Debate* between Shapley and Heber Curtis [3]. It was only in 1924, when Edwin Hubble identified Cepheids stars in the Andromeda *nebula* determining their distance, that the controversy could be definitively settled in favour of the island theory. Hubble proved, in fact, that the Andromeda nebula was much too far away to be part of our Galaxy and, therefore, that it should be regarded as a galaxy on its own. The distances of many other galaxies around the Milky Way were then measured and the structure of the so-called *Local Group* was reconstructed with a reasonable accuracy.

However, it was thought for long time that the Local Group should still occupy a central position in the universe. It was only in 1951 that the German astronomer Walter Baade could measure the distances of hundreds of galaxies outside the Local Group, showing that our Galaxy and the Local Group are pretty standard objects in a much bigger universe without any particularly privileged position.

These observations all supported the *cosmological principle* , the idea that, on sufficiently large scales, our universe has no privileged points or directions and that, therefore, it looks pretty much the same independently of the location of the observer and the direction of observation.[1]

It received even stronger support with the discovery of the *cosmic microwave*

---

[1]It is a stronger version of the Copernican principle since it assumes also isotropy in addition

background (CMB) radiation in 1965 by Arno Penzias and Robert Wilson of Bell Laboratories [5]. More recently, high redshift galactic surveys encompassing larger and larger portions of the observable universe have provided conclusive evidence.

In 1917 Einstein, starting from the perfect cosmological principle as a simplifying working hypothesis, derived the first cosmological solution within general relativity: Einstein's static model [6]. As shown by Eddington, this model was highly unstable. In 1922 Friedmann showed that, when the cosmological principle is assumed, Einstein's equations of general relativity reduce to a very simple set of equations that today bear his name: the *Friedmann equations* [7]. A few years later, in 1927, Lemaitre derived, on pure theoretical grounds, an expansion law [8] that two years later would be confirmed experimentally by Edwin Hubble and that bears today his name: *Hubbles's law* [9].

Later on, in 1935-36, Robertson and Walker managed to prove and write down the most general metric for a homogeneous and isotropic universe, i.e., respecting the cosmological principle. Moreover from the solutions of the Friedmann equations, Lemaitre was able to identify a particularly intriguing scenario where, going back in a finite interval of time, the universe should have been infinitely dense. According to this scenario the history of the universe would have started with initial conditions very different from those observed today, a scenario today popularly known as the *Big Bang theory*.[2] As we will discuss at length, this idea is today supported by a host of cosmological observations.

Another important breakthrough in cosmology took place when it became clear that the physics of the early universe is tightly connected to microphysics. It is then possible to reconstruct the history of the early universe by inserting our knowledge of nuclear and elementary particle physics in the cosmological picture. This was the beginning of *particle cosmology*, one of the most beautiful examples of synergetic scientific field. Its first great triumph is represented by the model of *Big Bang nucleosynthesis*, pioneered by George Gamow and collaborators at the end of the 1940s [10]. They showed that assuming the so-called *Hot Big Bang model* and using our knowledge of nuclear physics and statistical mechanics, it is possible to reproduce the observed primordial nuclear abundances. Twenty years later, in 1966, Yakov Zel'dovich, in collaboration with Semjon S. Gershtein, made another important step forward. They showed that, from cosmological considerations, it was possible to draw conclusions on yet laboratory untested elementary particle properties. Specifically they managed to place an upper bound on the neutrino masses [11]. In this way they showed that with particle cosmology, one cannot only reconstruct the history of the universe from the existing knowledge of fundamental

---

to homogeneity. It was first envisaged by Isaac Newton in his *Philosophiae Naturalis Principia Mathematica* (1687) [4].

[2]The name was coined in 1949 by Fred Hoyle in a pejorative way, proposer together with Bondi and Gold of the main competitor idea, the steady state theory. The original proposal had been made by George Lemaitre many years before, in 1931, who thought that all the universe could at the beginning be described as a primeval unstable atom whose radioactive decay could explain the start of the expansion.

physics, but also, reversing the logical direction of the connection, from cosmological observations one can derive new information on fundamental physics.

One year later, in 1967, Andrej Sakharov[3] made a new important step forward, showing how cosmological considerations could be used to test models of *new physics*, i.e., models beyond the standard model and general relativity. Inspired by the discovery made three years before by Cronin and Fitch that neutral kaons K_0 exhibit a different behaviour than their anti-particles $\bar{K}_0$ violating CP symmetry, he realised that this could be the origin of the observed matter-antimatter asymmetry of the universe. He proposed that the asymmetry could be generated dynamically during the early universe with a mechanism of what is now called *baryogenesis* relying on new physics beyond the standard model [3]. In this way, for the first time, cosmology was used as a special phenomenological tool for the investigation of new models of particle physics.

In more recent years, starting from the end of the 1970s, other important evidence of new physics from cosmology was discovered. It was gradually realised that it is necessary to postulate the existence of a hitherto undiscovered elementary particle to solve the so-called *dark matter puzzle*, that was first discovered in the study of stellar and galactic dynamics [13].

Baryogenesis and dark matter are not the only cosmological strong evidences of new physics. The necessity of a period of *inflation* at the beginning of the early universe history [14] and the necessity of introducing a new form of energy, commonly dubbed *dark energy*, to explain the acceleration of the universe expansion at present, are also considered as cosmological evidence for the existence of new physics.

In this way, with particle cosmology, the early universe becomes a very special laboratory of particle physics with far-seeing implications. These need to be contrasted with the results from conventional ground laboratories, for example particle colliders like the Large Hadron Collider (LHC) currently operating at the CERN laboratory. This approach will play a central role in our discussion, especially in the second part.

BIBLIOGRAPHY

[1] F. W. Bessel, *On the parallax of 61 Cygni*, Mon. Not. Roy. Astron. Soc. **4** (1838) 152.

[2] H. Shapley, *Studies based on the colors and magnitudes in stellar clusters. XII. Remarks on the arrangement of the sidereal universe*, Astroph. J. **49** (1919) 311.

[3] H. Shapley and H. D. Curtis, *The Scale of the Universe*, Bulletin of the National Research Council, Vol. 2, Part 3, No. 11, p. 171-217 (1921).

---

[3]Yes, the same Sakharov who was awarded the Nobel Prize for peace in 1975.

[4] Newton, Isaac, *Philosophiae Naturalis Principia Mathematica (Mathematical Principles of Natural Philosophy)*, London, 1687; Cambridge, 1713; London, 1726. (Pirated versions of the 1713 edition were also published in Amsterdam in 1714 and 1723.)

[5] A. A. Penzias and R. W. Wilson, *A measurement of excess antenna temperature at 4080-Mc/s*, Astrophys. J. **142** (1965) 419.

[6] A. Einstein, *Cosmological considerations in the general theory of relativity*, Sitzungsber. Preuss. Akad. Wiss. Berlin (Math. Phys.) **1917** (1917) 142.

[7] A. Friedman, *On the curvature of space*, Z. Phys. **10** (1922) 377 [Gen. Rel. Grav. **31** (1999) 1991].

[8] G. Lemaitre, *A homogeneous universe of constant mass and growing radius accounting for the radial velocity of extragalactic nebulae*, Annales Soc. Sci. Brux. Ser. I Sci. Math. Astron. Phys. A **47** (1927) 49.

[9] E. Hubble, *A relation between distance and radial velocity among extra-galactic nebulae*, Proc. Nat. Acad. Sci. **15** (1929) 168. doi:10.1073/pnas.15.3.168.

[10] G. Gamow, *Expanding universe and the origin of elements*, Phys. Rev. **70** (1946) 572; R. A. Alpher, H. Bethe and G. Gamow, *The origin of chemical elements*, Phys. Rev. **73** (1948) 803; R. A. Alpher and R. C. Herman, *On the relative abundance of the elements*, Phys. Rev. **74** (1948) 1737.

[11] S. S. Gershtein and Y. B. Zeldovich, *Rest mass of muonic neutrino and cosmology*, JETP Lett. **4** (1966) 120 [Pisma Zh. Eksp. Teor. Fiz. **4** (1966) 174].

[12] A. D. Sakharov, *Violation of CP invariance, C asymmetry, and baryon asymmetry of the universe*, Pisma Zh. Eksp. Teor. Fiz. **5** (1967) 32 [JETP Lett. **5** (1967) 24] [Sov. Phys. Usp. **34** (1991) 392] [Usp. Fiz. Nauk **161** (1991) 61].

[13] J. C. Kapteyn, *First attempt at a theory of the arrangement and motion of the sidereal system*, Astrophys. J. **55** (1922) 302; J.H. Oort, *The force exerted by the stellar system in the direction perpendicular to the galactic plane and some related problems*, Bull. Astron. Inst. Netherlands **6** (1932) 249; F. Zwicky, *Die Rotverschiebung von extragalaktischen Nebeln*, Helv. Phys. Acta **6** (1933) 110 [Gen. Rel. Grav. **41** (2009) 207]; V. C. Rubin and W. K. Ford, Jr., *Rotation of the Andromeda nebula from a spectroscopic survey of emission regions*, Astrophys. J. **159** (1970) 379.

[14] A. A. Starobinsky, *A new type of isotropic cosmological models without singularity*, Phys. Lett. **91B** (1980) 99; A. H. Guth, *The inflationary universe: a possible solution to the horizon and flatness problems*, Phys. Rev. D **23** (1981) 347.

CHAPTER 2

Fundamental observations

In this chapter we first introduce useful cosmological units of measurement and then we discuss those fundamental cosmological observations, the pillars of cosmology, that a model should be able to address in a unified picture, leading ideally to new phenomenological and testable predictions. Actually, it should be appreciated how, in most cases, the experimental discoveries were to some extent envisaged on pure theoretical grounds.

2.1 UNITS OF MEASUREMENT

Before giving an overview of the current cosmological observations, it will prove useful to introduce some convenient units of measurement and conventions.

Let us start by discussing *units of distance*. On astronomical scales, the meter is clearly not a very convenient unit. On planetary scales, it is convenient and customary to use the *astronomical unit* (symbol AU), defined as the mean distance between the Earth and the Sun: $1\,\mathrm{AU} \simeq 1.5 \times 10^{11}\,\mathrm{m} = 150 \times 10^{6}\,\mathrm{km}$. On cosmological scales, even the AU is too inconveniently small. A typical distance unit used to indicate large astronomical distances is the *light year* (ly), defined as the distance travelled by light in a vacuum in one year: $1\,\mathrm{ly} \simeq 1 \times 10^{16}\,\mathrm{m} \simeq 10{,}000 \times 10^{9}\,\mathrm{km}$.

However, for reasons that will be clear soon, the most commonly used unit of distance in astronomy is the *parsec* (symbol pc), the typical distance between neighbouring stars. It is defined as the distance at which 1 AU subtends a second of arc,[1]

$$1\,\mathrm{pc} \simeq 3.0856 \times 10^{16}\,\mathrm{m} \simeq 30{,}000 \times 10^{9}\,\mathrm{km} \simeq 3\,\mathrm{ly}\,. \tag{2.1}$$

On the other hand, the average distance between galaxies is much higher and is more conveniently expressed in terms of megaparsecs,

$$1\,\mathrm{Mpc} \equiv 10^{6}\,\mathrm{pc} \simeq 3 \times 10^{22}\,\mathrm{m}\,. \tag{2.2}$$

For example, the galaxy of Andromeda is approximately $0.7\,\mathrm{Mpc}$ far away from

---

[1]For comparison consider that 1 arcsecond is approximately the angle subtended by a coin at a distance of 1 mile.

us and it is the only extra-galactic object, barring the Magellanic clouds that are satellite dwarf galaxies of the Milky Way Galaxy, visible with the naked eye (it is located in the Andromeda constellation and it appears as a magnitude 4 nebula: the famous object M31 in the Messier catalogue).

It will be convenient to measure the *cosmological time* in billions of years (gigayears) and it will prove useful to have in mind the simple conversion relation

$$1\,\mathrm{Gyr} \simeq 3.16 \times 10^{16}\,\mathrm{s}\,. \tag{2.3}$$

Let us now briefly discuss *units of mass and energy*. The most convenient adopted unit of mass, both in astronomy and in cosmology, is the *solar mass*,

$$1\,M_{\odot} \simeq 2.0 \times 10^{30}\,\mathrm{kg}\,. \tag{2.4}$$

Our sun is also used as a sample for a convenient cosmological unit of luminosity, the *solar luminosity*,

$$1\,L_{\odot} \simeq 3.8 \times 10^{26}\,\mathrm{W}\,. \tag{2.5}$$

The most common unit of energy is the *electronvolt*,[2] defined as the kinetic energy acquired by an electron passing through a potential difference of 1 volt. Since the electron charge, in absolute value, is given by $e \simeq 1.6 \times 10^{-16}\,\mathrm{C}$, one correspondingly has

$$1\,\mathrm{eV} \simeq 1.6 \times 10^{-19}\,\mathrm{J}\,. \tag{2.6}$$

Since $[E] = [m\,c^2]$, one has a useful conversion relation between electronvolts and kilograms, given by

$$1\,\mathrm{eV}/c^2 \simeq 1.8 \times 10^{-36}\,\mathrm{kg}\,. \tag{2.7}$$

In cosmology, in particular during the early universe, temperature T plays a crucial role. In the SI, energy is obtained simply multiplying temperature by the Boltzmann constant k_B given by [3]

$$k_B \simeq 0.8617 \times 10^{-4}\,\mathrm{eV\,K^{-1}}\,. \tag{2.8}$$

A very useful numerical relation, allowing to express the reduced Planck constant $\hbar$ in terms of electronvolts and meters, is given by

$$\hbar\,c \simeq 0.19733\,\mathrm{GeV} \times 10^{-15}\,\mathrm{m}\,, \tag{2.9}$$

---

[2]This could at first sight appear as a too tiny unit of energy for cosmological needs but, as we will see, microscopic properties of matter and radiation, affect global properties of the universe, a key idea in *particle cosmology* that we will discuss in the second part on the early universe.

[3]The physical meaning of $k_B\,T$ comes from the definition of the Boltzmann constant in the statistical definition of entropy. Remember that in thermodynamics internal energy and temperature are related by the thermodynamic relation $dU = T\,dS$, valid for an infinitesimal thermodynamic transformation at constant volume and number of particles. From the statistical definition of entropy, $S = k_B \ln W$, where W is the number of quantum states of the accessible region of phase space, one immediately finds $k_B\,T = dU/dW$.

where $1\,\text{MeV} \equiv 10^6\,\text{eV}$ (megaelectronvolt) and $1\,\text{fm} \equiv 10^{-15}\,\text{m}$ (femtometer or fermi).

Another important quantity, representing historically the first attempt of combining together quantum mechanics and gravity, is the *Planck mass* defined as

$$M_P \equiv \sqrt{\frac{\hbar c}{G}} \simeq 1.2 \times 10^{19}\,\text{GeV}/c^2\,, \tag{2.10}$$

This allows to express the gravitational constant G in terms of the Planck mass, as

$$G = \frac{\hbar c}{M_P^2} \simeq 6.7 \times 10^{-38}\,(\text{GeV}/c^2)^{-2}\,\hbar c\,. \tag{2.11}$$

The Planck mass is the mass of two identical point-like particles whose absolute value of the gravitational potential energy times their distance is equal to $\hbar c$. For larger masses one expects quantum effects to play a role in gravity, since there is a clear violation of the uncertainty principle.

Using Eqs. (2.1) and (2.4), we can also find an expression of the value of the gravitational constant, that will prove convenient when discussing the dynamics of clusters of galaxies, in solar masses and megaparsecs:

$$G \simeq 4.3 \times 10^{-9}\,\text{km}^2\,\text{s}^{-2}\,\text{Mpc}\,M_\odot^{-1}\,. \tag{2.12}$$

In the *natural system of units* $k_B = \hbar = c = 1$. It is not only a practical tool to simplify equations, but it also highlights important fundamental connections among physical quantities and for this reason it is widely used in theoretical physics.

Units of temperature, mass, energy, length and time can be all expressed in terms of just one independent unit of a physical quantity, typically energy.[4] In particular, since $k_B = 1$, temperature has the dimension of energy and can be measured in electronvolts. Kelvin degrees and electronvolts are then simply related by the conversion factor

$$1\,\text{K} \simeq 0.8617 \times 10^{-4}\,\text{eV}\,. \tag{2.13}$$

Since $c = 1$, in natural units, energy, mass and momentum have all the same dimension and can be measured in electronvolts. This means that kilograms and electronvolts have the same dimensionality and Eq. (2.7) becomes simply a conversion equation: $1\,\text{eV} \simeq 1.8 \times 10^{-36}\,\text{kg}$.

In particular, the numerical equality Eq. (2.10) for the Planck mass simply becomes $M_P \simeq 1.2 \times 10^{19}\,\text{GeV}$, the highest known energy scale in physics. Notice also that Eq. (2.11), expressing the gravitational constant in terms of the Planck mass, simply becomes $G = 1/M_P^2 \simeq 6.7 \times 10^{-38}\,\text{GeV}^{-2}$, showing how the weakness of the gravitational force can be interpreted in terms of the largeness of the Planck energy scale and vice versa.

In natural units, from Eq. (2.9), one simply obtains the conversion relation

---

[4]In this case in natural units one can just use electronvolts and its multiples as units of measurement.

$1\,\mathrm{GeV} \simeq 5\,\mathrm{fm}^{-1}$, showing how the energy has the same dimension of an inverse length and vice versa. This is basically a consequence of the uncertainty principle, that has a very important phenomenological implication in elementary particle physics: in order to probe physics laws at smaller distances, we need to go to higher energies, a basic aspect in collider physics.

Throughout the book we will adopt the SI, using natural units only for temperature ($k_B = 1$).[5] Therefore, we will still indicate c and $\hbar$ in the expressions. This will make it easier for the reader to derive useful numerical values of the physical quantities in SI units, especially when we will deal with cosmological distances and times.

2.2 NOT JUST ELECTROMAGNETIC RADIATION

How do we observe the universe? Mainly by means of the electromagnetic radiation emitted by different kinds of luminous sources. For example we can observe galaxies in visible light and the cosmic microwave background (CMB) radiation in the microwaves.

Indeed the electromagnetic radiation offers an extensive variety of options and corresponding methods of observation depending on the various frequency ranges:

- Radio-waves (radio-astronomy);

- Microwaves (microwave astronomy);

- Infrared (infrared astronomy);

- Visible (visible-light astronomy);

- Ultraviolet (UV astronomy);

- X-rays (X-ray astronomy);

- γ-rays (γ astronomy).

However, there are also quite interesting, more modern, additional possibilities as exploratory probes beyond the electromagnetic radiation. For example, the most energetic component of the cosmic rays, made of different kinds of particles like protons, nuclei, electrons, muons, neutrinos and other elementary particles, could be of cosmological origin.

Neutrinos in particular are a promising way to get cosmological information.

---

[5]This is equivalent to define entropy in a dimensionless way: $S = \ln W$. It is literally the natural definition of temperature and entropy and one can fully replace Kelvin degrees with electronvolts harmlessly. The use of $c = \hbar = 1$ requires more advanced training, since in many situations it is convenient to express scales of time and length in SI units, rather than in multiples of electronvolts and switching from natural units to SI units requires a certain fluency in recovering correctly $\hbar$ and c factors. To this extent, when possible, we always express SI units in units of c, $\hbar$ and their combinations, providing basically a dictionary between the two systems. For example, we will usually express masses in eV/c^2 rather than kilograms.

At relatively low energies $\mathcal{O}(10\,\mathrm{MeV})$ neutrinos from supernovae explosions at astronomical distances can be detected. Actually 11 neutrinos from the SN 1987a have been indeed observed in 1987 by the Kamiokande neutrino detector in Japan. These neutrinos from supernovae explosions can give us very precious information not only on the mechanism of the supernovae explosion itself, but potentially they could also be used to obtain interesting cosmological information. Even more interestingly for cosmological applications, in April 2013, neutrinos of cosmological origin, with very high energies in the range $10\,\mathrm{TeV}$–$10\,\mathrm{PeV}$, have been observed in the IceCube detector located at the South Pole [1], marking the beginning of a new exciting step in *neutrino astronomy*[6] with potentially important cosmological implications.

In 2009 the PAMELA satellite detected an excess of positrons[7] in cosmic rays with respect to theoretical expectations [2]. Such an anomaly has been interpreted as a possible signature of those still unknown particles that would constitute the so-called dark matter of the universe. However, more recent data seem to support a more astrophysical explanation of such a positron excess. Nevertheless, future cosmic rays measurements could indeed finally provide the first direct (non-gravitational) evidence of the existence of dark matter particles.

The existence of gravitational radiation (or gravitational waves) was first predicted by Einstein as a consequence of the theory of general relativity in 1916. Since then they have been intensively searched as the last untested[8] prediction of general relativity. The discovery of gravitational waves made by the combined LIGO and Virgo collaborations [3] analysing data taken by the two LIGO laser interferometer gravitational detectors in September 2015 and announced in February 2016,[9] marked the birth of a new kind of astronomy with an expected huge impact in astrophysics during future years. However, the discovery of gravitational waves from astrophysical sources also gives hope for the (certainly more challenging) direct detection of relic cosmological gravitational waves.[10] These are a kind of cosmological holy grail, since they might potentially probe very early stages in the history of the universe, much earlier than the epoch probed by the CMB radiation. There are different mechanisms predicting the emission of gravitational radiation just a tiny fraction of a second after the Big Bang with some potential to be detected.

---

[6]Detectors of neutrinos from space are also sometimes informally called *neutrino telescopes*.

[7]Positrons are the anti particles of electrons. They have the same mass but opposite electric charge, i.e., positive.

[8]Indirectly they were proved to exist in 1974 thanks to the gravitational radiation emitted in a close binary system, the Hulse–Taylor binary, causing their orbit to shrink and their orbital period to decrease. For this discovery, R.A. Hulse and J.H. Taylor were awarded the Nobel Prize in 1993.

[9]This was possible detecting a signal unambiguously identified as the signature of gravitational waves emitted by a binary black hole system merger about 1.3 billion years ago at a distance of 410 Mpc. For this discovery the Nobel Prize for Physics 2017 has been awarded to R. Weiss, B. C. Barish and K.S. Thorne, *for decisive contributions to the LIGO experiment and the observation of gravitational waves.*

[10]Indirectly, they might have left an imprint even in the CMB temperature anisotropies.

2.3 MEASURING DISTANCES: THE COSMIC DISTANCE LADDER

There are different ways to measure the distance of an electromagnetic radiation astronomical source.

2.3.1 Parallax distance

Suppose we observe, from a point O, a luminous source in the sky located at a distance d_π. Suppose also we are able to observe again the same luminous source from a second point P located at a distance b in a direction perpendicular to the line of sight from O. The angular position of the object on the sky would then change by an angle $\theta = \arctan(b/d_\pi)$ called *parallax*. Vice versa, the *parallax distance* of an object is obtained as

$$d_\pi \equiv \frac{b}{\tan \theta} \simeq \frac{b}{\theta} = 1\,\mathrm{pc}\,\frac{b}{1\,\mathrm{AU}}\,\frac{1\,\mathrm{arcsec}}{\theta}, \tag{2.14}$$

where we made a small angle approximation $\theta \ll 1$. Notice that for $b = 1\,\mathrm{AU}$ one obtains $d_\pi = 1\,\mathrm{pc}$, in agreement with the definition of parsec. Typically one measures the *annual parallax*, due to the motion of the Earth around the Sun, and in this case $b = 1\,\mathrm{AU}$.[11] Historically, as mentioned, the first annual parallax to be measured was that one of the star 61 Cygni, with $\theta \simeq 0.3\,\mathrm{arcsec}$ corresponding to a distance $d_\pi \simeq 3\,\mathrm{pc}$. Note that, though it is among closest stars with highest annual parallax, it clearly respects the small angle approximation: that is why it took thousands of years for this effect to be measured after its prediction!

2.3.2 Luminosity distance

A luminous astronomical source with a known *absolute luminosity* L, the total emitted power, is called *standard candle*. Assuming that a simple inverse-square law for the reduction of the light intensity with distance holds, the power received by a detector of area A located at a distance d_L is given by

$$P = L\,\frac{A}{4\,\pi\,d_L^2}. \tag{2.15}$$

Knowing L and measuring P, one can derive the *luminosity distance* defined as

$$d_L \equiv \left(\frac{L}{4\,\pi\,F}\right)^{1/2}, \tag{2.16}$$

where $F \equiv P/A$ is the measured flux, i.e., the power received per unit area. This definition can be extended to a more general situation where the given assumptions, inverse square law and fixed distance, do not necessarily hold.

---

[11]The annual parallax is defined as the angular shift of a star due to the annual motion of the Earth around the Sun of a distance equal to the average Earth-Sun distance, i.e., 1 AU, along a direction orthogonal to the line-of-sight of the star. It is of course also corresponding to the angle subtended by the astronomical unit at the distance of the star.

2.3.3 Angular distance

Finally, suppose we know the length ℓ of an object called in this case *standard yardstick*. Measuring its angular size $\delta\theta \ll 1$, one can derive the *angular distance* defined as

$$d_A \equiv \frac{\ell}{\delta\theta}. \tag{2.17}$$

In a Euclidean space with fixed distances among the objects, the various distances coincide with each other, but as we will see on cosmological scales these assumptions do not hold in general, in particular for quite large distances above approximately 100 Mpc. As we will see, a comparison among the measured values of the different distances, can be used as a probe to test different cosmological models and to determine the relevant cosmological parameters.

2.3.4 Cosmic distance ladder

Let us now see how it was possible in practice, during the last 100 years, to draw a map of the astronomical objects around us going much beyond the borders of the Local Group.

A precise determination of the astronomical unit is the first step. Today this is done with great precision with techniques based on radar reflection off the surfaces of planets. This determination can be regarded as a first rung in the so-called *cosmic distance ladder*.

As we discussed, by measuring the stellar annual parallaxes of sufficiently close stars, one can then determine their parallax distance (see Eq. (2.14)). One cannot determine stellar parallaxes lower than ~ 0.01 arcsec and therefore this method can be used to determine star distances within ~ 100 pc around us. Notice that a precise determination of the parsec also relies on a precise determination of the AU. This specific example explains the use of the word ladder: each step up to a higher rung depends on the calibration made using the method covering shorter distances with the effect that errors from all previous steps propagate to the next one. Therefore, it is crucial to find objects whose distance can be determined with two different methods. These objects can then be used as calibrators (so-called *distance anchors*) for the higher rung in the cosmic ladder.

For distances beyond the reach of the annual parallax method, the difficulty is to be able to know the absolute luminosity L or the length ℓ of an object. In other words the main problem is the identification of standard candles and/or standard yardsticks.

One can use the annual parallax in order to calibrate standard candles located within the reach of the annual parallax method. Most of the stars can be used as standard candles if they are close enough to allow a determination of their spectral class (stars are classified in spectral classes on the basis of features observed in their spectrum, due for example to their temperature, size and chemical composition). If one knows the distance of a star, determined through its annual parallax, then the absolute luminosity can be calculated using $L = 4\pi d^2 F$. In this way it has been discovered that there is a relation between the luminosity of a star and the spectral type, the famous Hertzsprung–Russell (H-R) diagram. One can then apply this relation in reverse to stars at a distance beyond the reach of the annual parallax method. This means that measuring their spectral class and using the H-R relation one can deduce the absolute luminosity and from this calculate the luminosity distance. This method works particularly well for clusters of stars (globular clusters or open clusters), where stars have approximately the same distance. As first pointed out by Walter Baade, one has however to be careful, distinguishing the H-R diagram of stars of Population I (younger) and that of stars of Population II (older). Using the H-R diagram it has been possible to determine the distance of stars up to 100 kpc.

A very important class of standard candles is given by variable stars for which there is an empirical evidence of a relation between their *luminosity curves* and in particular the period of variability and their absolute luminosity. There are three famous kinds of such variables:

- *RR* Lyrae, with a period between a few hours to one day. They allow to determine distances up to ~ 1 Mpc;

- *Cepheids*, with a period ranging from 2 to 40 days; they allow to determine distances up to ~ 10 Mpc. In particular in 1924 the discovery of a Cepheid made by Hubble in the Andromeda Galaxy made it possible to determine the distance of the Andromeda galaxy hosting the Cepheid (about 0.7 Mpc). This distance is much bigger than the size of our galaxy. In this way it became possible to give a solution to quite a strong scientific controversy, whether the Milky Way Galaxy should be considered encompassing the whole universe or just one among billions of other galaxies. Cepheids also played a crucial role in the discovery of the universe expansion made by Hubble in 1929. In 1994 the Hubble Space Telescope managed for the first time to determine the distance of a Cepheid in the Virgo Cluster about ~ 17 Mpc from us. This made possible a precise calibration of the next rung in the cosmic distance ladder and the most precise measurement of the Hubble constant, the parameter that describes the expansion of the universe at present, before 2013, when the *Planck* satellite provided the most precise measurement of the Hubble constant from CMB observations.

- *Supernovae type Ia*. They have a very well determined maximum luminosity in their light curve, the change of luminosity with time after the explosion. They are useful in determining the distances of the furthest galaxies, up to a few hundred Mpc. In 1998 they surprisingly provided the first strong observational evidence of an accelerated expansion of the universe at present.

Another important class of standard candles is represented by *spiral galaxies*, since their luminosity is strongly correlated with their maximum rotation velocity (*Tully–Fisher relation*). This method works for distances up to ~ 100 Mpc.

An example of *standard yardstick* is given by *eclipsing binaries* stars. One can reconstruct the orbital parameters of these binary systems and in particular the size of the orbit. With this technique it is possible to measure distances up to 3 Mpc. Therefore they are useful extra-galactic standard yardsticks.

Thanks to the calibration of the cosmic distance ladder, astronomical observations have been able to reconstruct the large-scale structure of the universe around us, one of the greatest achievements of modern astronomy.

2.3.5 Standard sirens

The discovery of gravitational waves paves the way to a completely new method to measure the distance of an object that does not rely on the cosmic distance ladder and, therefore, offers a completely independent cross check. The gravitational wave frequency modulation, the so-called *chirp*, generated by the merging of a compact binary system, also contains information on the mass of the emitting object and from this one can calculate the absolute amplitude of the signal. When this is compared with the measured amplitude, since one knows that this decays inversely proportionally with the distance of the observer from the emitting source, one can immediately extract the distance of the source. Such objects have been dubbed *standard sirens* [4].

2.4 COSMOLOGICAL PRINCIPLE

The observations of how galaxies are distributed around us, the so-called *large-scale structure of the universe*, show that there are no structures in the universe bigger than galaxy super-clusters, with a size ~ 100 Mpc. Therefore, on larger scales the universe appears approximately homogeneous, i.e., it presents the same physical properties at any point.[12] At the same time the observations do not show any evidence of the existence of a privileged direction around us (isotropy) and therefore, if homogeneity holds, around any point. The homogeneity and isotropy of the universe on large scales is also well supported by the almost uniform temperature angular distribution of the CMB on the sky. As we discussed in Chapter 1, the simultaneous assumption of homogeneity and isotropy is referred to as cosmological principle.

---

[12]When we say that we observe no structure in the universe on scales larger than 100 Mpc, we imply an operation of smearing-out on a sphere of radius ~ 100 Mpc centred around each point. More in detail, such a smearing-out operation consists in replacing the values of intensive physical quantities at each point with their mean values on a sphere with a radius R. For example for the matter density one has to replace

$$\rho(\vec{r}) \to \overline{\rho}_R(\vec{r}) = \frac{\int_{S(\vec{r},R)} d^3 r' \, \rho(\vec{r'})}{(4\pi/3) \, R^3} . \tag{2.18}$$

Therefore, when we say that the universe is homogeneous on scales larger than 100 Mpc it means that for $R \gtrsim 100$ Mpc fluctuations $\delta\rho(\vec{r}) \equiv [\overline{\rho}_R(\vec{r}) - \overline{\rho}_{R\to\infty}]/\overline{\rho}_{R\to\infty} \ll 1$. The cosmological principle applies to such a smeared-out universe. This is not in contradiction to the existence of inhomogeneities on scales much smaller than 100 Mpc.

From Einstein's equations, imposing the cosmological principle, one derives the Friedmann equations, describing the dynamics of the cosmological gravitational field in a homogeneous and isotropic universe. This defines a class of models called Friedmann (or Friedmann–Lemaitre) cosmological models (or sometimes simply Friedmann cosmology). As we will see, different models are obtained depending on the assumption on the nature of the fluids filling the universe.

A clear important issue that one has to address, is whether time homogeneity applies as well or in other words whether the universe can be considered static or if it dynamically evolves.

2.5 COSMOLOGICAL EXPANSION AND HUBBLE'S LAW

Consider a luminous signal emitted by an astronomical object with wavelength λ_{em} and detected by an observer, at the present time, with wavelength λ_0.[13] The *redshift* is defined as

$$z \equiv \frac{\lambda_0 - \lambda_{\text{em}}}{\lambda_{\text{em}}} . \qquad (2.19)$$

Hubble found that galactic redshifts follow approximately the relation (*Hubble's law*)

$$z \simeq \frac{H_0 \, d_L}{c} , \qquad (2.20)$$

where d_L is the luminosity distance of the galaxy from us. The quantity H_0 is called *Hubble constant*, since it has the same value for all galaxies independently of their direction of observation and distance.

Interpreting the redshift as a non-relativistic Doppler effect one has $z = v/c$, where v is the radial velocity of the object. In this case one concludes that galaxies are receding from us with a velocity proportional to their distance, namely

$$v \simeq H_0 \, d_L . \qquad (2.21)$$

One can see from the left panel of Fig. 2.1 that, in his original observations, Hubble could not observe galaxies further than a few megaparsecs and in this case there is a large dispersion of the galactic velocities around the best fit values from Hubble's law, that, therefore, has to be regarded more as a sort of intuitive guess. On the other hand, with the Hubble Space Telescope (HST), it has been possible in the last few years to determine the distance of much further objects, up to distances of a few hundred megaparsecs. One can see in the right panel of Fig. 2.1 that the dispersion of the galaxy velocities, around the value dictated by Hubble's law for a given distance, is much smaller. This is because galaxies are moving with respect to us with a velocity given by the sum of the Hubble velocity (that is always a recession velocity) plus what is called the peculiar velocity of the i-th galaxy that

---

[13]More generally, through all the notes, and as customary in cosmology, all quantities at the present time will be indicated with the subscript "0".

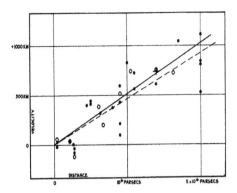
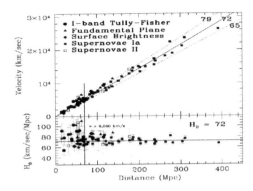

FIGURE 2.1: Comparison of the plots of velocity-distance used to determine the Hubble constant: on the left the original plot made by Hubble using 22 *extragalactic nebulae* at distances up to 2 megaparsecs [5], on the right the plot obtained by the HST collaboration plot using the indicated objects at distances up to 400 megaparsecs [6]. The velocity has to be meant simply the quantity that would correspond to non-relativistic Doppler effect $v = c\,z$ and it is then just a different way to plot the redshift.

we indicate with $\vec{v}_{p,i}$. Explicitly we can, therefore, express the velocity of the i-th galaxy as

$$\vec{v}_i \simeq H_0\,\vec{r}_i + \vec{v}_{p,i}\,. \tag{2.22}$$

This can have any orientation, and therefore it can be either a recession velocity $(\vec{v}_{p,i}\cdot\vec{r}_i > 0)$ or an approaching velocity $(\vec{v}_{p,i}\cdot\vec{r}_i < 0)$. However, while the magnitude of the Hubble velocities increases linearly with the distance, the magnitude of the peculiar velocities is not higher than $\sim 1000\,\mathrm{km/s}$. Therefore, sufficiently distant galaxies recede away from us with speeds well described by Hubble's law. Therefore, for a precise measurement of the Hubble constant, one has to determine the distance of sufficiently far objects. However, we will see that if the objects are too distant, then further effects of a theoretical nature have to be taken into account. For this reason Hubble's law has to be regarded as an approximate relation holding averaging out peculiar velocities and for not too large distances.[14]

From the left panel of Fig. 2.1 one can also see how the original estimation of H_0 obtained by Hubble [5], $H_0 \sim 500\,\mathrm{km\,s^{-1}\,Mpc^{-1}}$, was a very poor one.[15] Indeed, it is almost 10 times higher than the current measurement from the Hubble Space Telescope (HST) [6]

$$H_0 = (72 \pm 8)\,\mathrm{km\,s^{-1}\,Mpc^{-1}}\,. \tag{2.23}$$

---

[14]We will rigorously specify what 'not too large distances' means.

[15]The large overestimation of H_0 made by Hubble originated from the fact that at that time the different calibration of luminosity curves needed for Cepheids of Population I and Population II was still unknown and because in most distant galaxies he misinterpreted single stars as star clusters.

The HST determination of H_0 has been very recently further improved, reducing the statistical error to $\simeq 2.3\%$, [7]

$$H_0 = (73.48 \pm 1.66)\,\mathrm{km\,s^{-1}\,Mpc^{-1}}\,. \tag{2.24}$$

In March 2013, the *Planck* satellite collaboration obtained from CMB observations an independent very precise ($\simeq 1.5\%$) measurement [8],

$$H_0 = (67.3 \pm 1.2)\,\mathrm{km\,s^{-1}\,Mpc^{-1}}\,, \tag{2.25}$$

valid, as we will see, within the ΛCDM cosmological model. The ($3.7\,\sigma$) tension with the measurement from the HST in Eq. (2.24) is interpreted as the effect of some underestimated systematic uncertainty in the HST determination or, intriguingly, as a possible indication of the necessity to extend the ΛCDM model, for example introducing a new unknown ultra-relativistic contribution to the energy density at the time of recombination dubbed *dark radiation*.[16] This should be approximately equivalent to half of the contribution from a standard neutrino species. The latest measurement from the *Planck* satellite [9],

$$H_0 = (67.81 \pm 0.92)\,\mathrm{km\,s^{-1}\,Mpc^{-1}}\,, \tag{2.26}$$

slightly alleviates the tension.

It is interesting that using the recent combined detection of gravitational waves and γ-rays from a binary neutron star system, it was possible not only to measure the distance from the gravitational waves detection using the source as a standard siren, but also the redshift from the γ-rays detection. In this way it has been possible, following an earlier proposal [4], to derive a third completely independent measurement [10]

$$H_0 = (70^{+12}_{-8})\,\mathrm{km\,s^{-1}\,Mpc^{-1}}\,. \tag{2.27}$$

At the moment this measurement does not solve the tension between the two measurements, but this should be achieved with a few further additional detections of such systems that will reduce the error to $\sim 1\%$.

In the following we will use mainly the *Planck* result in Eq. (2.26) as reference value of H_0 for the determination of different quantities. However, because of the large experimental uncertainty that historically plagued the determination of H_0 and that seems to persist today to a lower extent, it is customary to parameterise H_0 in terms of a dimensionless parameter h (not to be confused with the Planck constant!) defined by

$$H_0 = 100\,h\,\mathrm{km\,s^{-1}\,Mpc^{-1}}\,. \tag{2.28}$$

Therefore, the *Planck* satellite result Eq. (2.26) can also be recast in terms of h as

$$h = 0.678 \pm 0.009\,. \tag{2.29}$$

---

[16]In cosmology any ultra-relativistic particle species can be regarded as 'radiation', since it approximately respects the equation of state of radiation $p = \varepsilon/3$, where p is the pressure and ε is the energy density.

From the Hubble constant it is also possible to define the *Hubble time* and the *Hubble radius* at present, respectively, as

$$t_{H,0} \equiv H_0^{-1} = (14.4 \pm 0.2)\,\text{Gyr}, \tag{2.30}$$

$$R_{H,0} \equiv c\,H_0^{-1} = (4.42 \pm 0.06)\,\text{Gpc}. \tag{2.31}$$

Where does Hubble's law come from? We will be able to answer this question in detail, on quantitative grounds, in Chapter 10, within Friedmann cosmological models. However, here it is useful to anticipate a qualitative discussion of some of the results that we will find, in particular on the issue of whether Hubble's law is compatible with the cosmological principle. At first sight it could seem that Hubble's law implies that we live in a very special point of observation, a kind of centre of the universe from which all galaxies are moving apart like fragments ejected by an explosion with spherical symmetry (we will discuss this kind of model in the next chapter). However, as we will see in the next chapter, there is a special kind of gas expansion that can explain Hubble's law without any necessity of assuming a special position of the observer but still assuming that there is indeed some special point in the universe from which the expansion originated, somehow at odds with the cosmological principle.

On the other hand, we will discover that there is another intriguing possibility offered by general relativity: the cosmological principle can be compatible with Hubble's law if one considers a model where all galaxies are moving apart from each other as if the whole space itself is stretching. In this sense the expansion of the universe should be regarded not as a traditional thermodynamical expansion of a gas in space but rather as an expansion of the space itself. Using a simple two-dimensional space analogy, one can think of points on the surface of an inflating spherical balloon. This model has the great advantage to predict immediately a linear relation between recession velocity and distance between objects. Conversely, as we will discuss in the next chapter, in a model where the cosmological expansion originated from some specific point, playing the role of centre of the universe, as the result of some initial explosion, Hubble's law does not derive from a physical model but has rather to be imposed. In an isotropic explosion one would indeed expect that the velocities of the ejected fragments decrease with the distance from the centre. Moreover this naive model does not have any chance to explain the CMB radiation observations and the observed high-redshift objects ($z \gtrsim 1$). Therefore, the popular name *Big Bang* is misleading and should not be meant literally as an explosion originated from a special point but rather as an explosion of the space itself.

2.6 CMB RADIATION

The furthest luminous electromagnetic radiation that we observe does not come from a compact luminous object but from all points in the sky. It is indeed a diffuse background called CMB radiation [11]. The spectrum of the CMB is one of the most precise realisations of a Planck distribution in Nature. The deviations of the observed CMB spectrum by a Planck spectrum with a temperature [12]

$$T_0 = (2.7255 \pm 0.0006)\,\text{K} \tag{2.32}$$

are below 0.01 %. As we will see, the CMB plays a crucial role in pinpointing the Hot Big Bang model as the right cosmological model.[17] Quantum-mechanically the CMB radiation can be regarded as a photon background. CMB photons are usually referred to as *relic photons*, since, as we will show in Chapter 12, they are the leftover of an early stage in the history of the universe. Their number and energy density can be quite easily derived from statistical mechanics.[18] Let us just highlight the most important points. Photons are particles with spin 1 and therefore, from the *spin-statistics theorem*,[19] they are bosons. Therefore, if thermal equilibrium applies to CMB photons, they obey the Bose–Einstein statistics and their distribution function is given by

$$f_{\gamma,0}(|\vec{p}|) = \frac{1}{e^{c\,|\vec{p}|/T_0} - 1}, \tag{2.33}$$

where T_0 is the temperature of the CMB at the present time.

The number density of relic photons can be easily calculated from the distribution function integrating over all quantum states obtaining

$$n_{\gamma,0} = \frac{2\,\zeta(3)}{\pi^2} \frac{T_0^3}{(\hbar\,c)^3} \simeq 410.7\,\text{cm}^{-3}. \tag{2.34}$$

where $g_\gamma = 2$ is the number of the spin degrees of freedom (or *spin degeneracy*) of photons,[20] $\zeta(x)$ is the ζ-Riemann function ($\zeta(3) = 1.20206$) and we used Eqs. (2.9), (2.13) and (2.32) to calculate the numerical value. One can proceed analogously for the calculation of the *energy density of relic photons* $\varepsilon_{\gamma,0}$. As an intermediate step one can also derive the *Planck spectrum*

$$\frac{d\varepsilon_{\gamma,0}}{d\nu} = \frac{16\,\pi^2\,\nu^3}{e^{\frac{2\,\pi\,\nu}{T_0}} - 1}, \tag{2.35}$$

fitting impressively well the observed spectrum with $T_0 = 2.7255\,K$. When this is integrated over frequency ν, one obtains

$$\varepsilon_{\gamma,0} = \frac{\pi^2}{15} \frac{T_0^4}{(\hbar\,c)^3} \simeq 0.2605\,\text{MeV}\,\text{m}^{-3}. \tag{2.36}$$

---

[17]In fact the existence of a fossil radiation had been first predicted within the Hot Big Bang model, much before its discovery, by Alpher and Hermann who even estimated its temperature to be $T \simeq 5°$ K [13].

[18]See exercise 2.8.

[19]The spin-statistics theorem states that integer spin particles are bosons and half-integer spin particles are fermions (e.g., electrons with $S = 1/2$ are fermions).

[20]While for all the other particles (except neutrinos as we will see) the spin degeneracy is given by $g_S = 2\,S + 1$, in the case of photons, having spin $S_\gamma = 1$, this does not apply since photons are massless and the longitudinal polarisation, corresponding to the spin oriented along the travelling direction, is missing.

2.7 PRIMORDIAL NUCLEAR ABUNDANCES

Nuclear abundances are synthesised in stars. In particular heavy nuclei like metals can be well explained within a traditional stellar nucleosynthesis. However, light elements like deuterium and ^{4}He, have an abundance that is in some cases much higher than what is predicted by stellar nucleosynthesis.

More precisely, the astronomical observations indicate that the universe is basically composed ∼ 75% hydrogen-1 (protium) and ∼ 25% ^{4}He, plus very small traces of other light elements like hydrogen-2 (deuterium), helium-3 and lithium-7. Only a small fraction of these nuclear abundances can be explained in terms of a stellar nucleosynthesis. As we will see the *Hot Big Bang model* will provide an elegant explanation of the primordial nuclear abundances with so-called Big Bang Nucleosynthesis.

2.8 LARGE-SCALE STRUCTURE OF THE UNIVERSE

On scales smaller than 100 Mpc the luminous matter is not homogeneously distributed but is organised in structures in a hierarchical way: stars form galaxies, galaxies form clusters of galaxies and finally these form super-clusters of galaxies, the biggest structures in the universe. In the first part we will basically neglect the existence of structures on small scales. This is indeed quite a technical subject in cosmology that will require different tools and that will present a clear challenge. We will see that the CMB radiation is the relic trace of an early stage of the universe called recombination which occurred only about 400,000 years after the Big Bang. The CMB radiation is, with very good approximation, homogeneous and isotropic, with temperature anisotropies in the sky at the level of 10^{-5} compared to the average temperature. It is then quite challenging for a cosmological model to explain how first galaxies formed only ∼ 350 million years after recombination. The presence of a so-called *cold dark matter* (CDM) component plays a crucial role in this respect within our current cosmological model.

EXERCISES

Exercise 2.1 *Knowing that 1 AU ≃ 1.5 × 10¹¹ m and that the speed of light c ≃ 3 × 10⁸ m, show respectively that 1 pc ≃ 3 × 10¹⁶ m and 1 ly ≃ 1 × 10¹⁶ m.*

Exercise 2.2 *The mass of a nucleon (a proton or a neutron) is approximately given by $m_N ≃ 1.7 × 10^{-27}$ kg. Find the corresponding value in eV/c^2.*

Exercise 2.3 *The typical mass of a galaxy is given by $10^{12} M_☉$. Assuming that nucleons contribute 1/5 to the mass of a galaxy, estimate the typical number of nucleons in a galaxy.*

Exercise 2.4 *Knowing that the gravitational constant is given by $G = 6.67 × 10^{-11}$ m³ kg⁻¹ s⁻², find the value of the Planck mass in GeV/c^2.*

Exercise 2.5 *Using the value for H_0 given in Eq. (2.26), verify the values of the Hubble time and of the Hubble radius in Eqs. (2.30), (2.31).*

Exercise 2.6 *Derive an expression for the frequency, as a function of temperature, corresponding to the peak of the Planckian distribution; calculate then the corresponding energy of photons of the CMB radiation with a frequency equal to the maximum of the Planckian distribution using $T_0 = 2.7255\,K$;*

Exercise 2.7 *How far should astronomical sources be in order to determine the Hubble constant with a precision lower than 10% using distance versus redshift data set?*

Exercise 2.8 *Starting from the Bose–Einstein distribution, derive the expressions (2.34) and (2.36) for the number and energy density of relic photons, respectively.*

Exercise 2.9 *Verify the numerical estimations in Eqs. (2.34) and (2.36) for the number and the energy density of relic photons, respectively. How many relic photons cross a human body every second (neglect atmospheric absorption)?*

BIBLIOGRAPHY

[1] M. G. Aartsen et al. [IceCube Collaboration], Phys. Rev. Lett. **111** (2013) 021103 [arXiv:1304.5356 [astro-ph.HE]].

[2] O. Adriani et al. [PAMELA Collaboration], *An anomalous positron abundance in cosmic rays with energies 1.5-100 GeV*, Nature **458** (2009) 607.

[3] B. P. Abbott et al. [LIGO Scientific and Virgo Collaborations], *Observation of gravitational waves from a binary black hole merger*, Phys. Rev. Lett. **116** (2016) 6, 061102 [arXiv:1602.03837 [gr-qc]].

[4] B. F. Schutz, *Determining the Hubble constant from gravitational wave observations*, Nature **323** (1986) 310; D. E. Holz and S. A. Hughes, *Using gravitational-wave standard sirens*, Astrophys. J. **629** (2005) 15.

[5] E. Hubble, *A relation between distance and radial velocity among extra-galactic nebulae*, Proc. Nat. Acad. Sci. **15** (1929) 168.

[6] W. L. Freedman et al. [HST Collaboration], *Final results from the Hubble space telescope key project to measure the Hubble constant*, Astrophys. J. **553** (2001) 47.

[7] A. G. Riess et al., *New parallaxes of galactic Cepheids from spatially scanning the Hubble space telescope: implications for the Hubble constant*, accepted by ApJ, [arXiv:1801.01120 [astro-ph.CO]].

[8] P. A. R. Ade et al. [Planck Collaboration], *Planck 2013 results. XVI. Cosmological parameters*, Astron. Astrophys. **571** (2014) A16

[9] P. A. R. Ade et al. [Planck Collaboration], *Planck 2015 results. XIII. Cosmological parameters*, arXiv:1502.01589 [astro-ph.CO].

[10] B. P. Abbott et al. [LIGO Scientific, Virgo, 1M2H, DLT40, Las Cumbres Observatory, VINROUGE and MASTER Collaborations], *A gravitational-wave standard siren measurement of the Hubble constant*, Nature [arXiv:1710.05835 [astro-ph.CO]].

[11] A. A. Penzias and R. W. Wilson, *A measurement of excess antenna temperature at 4080-Mc/s*, Astrophys. J. **142** (1965) 419.

[12] D. J. Fixsen, *The temperature of the cosmic microwave background*, Astrophys. J. **707** (2009) 511 [arXiv:0911.1955].

[13] R. A. Alpher and R. C. Herman, *On the relative abundance of the elements*, Phys. Rev. **74** (1948) 1737.

CHAPTER 3

A Newtonian cosmology?

In this chapter we attempt to build a cosmological model, aiming at explaining the fundamental cosmological observations discussed in the last chapter, within Newton's theory of gravity. We will conclude that a cosmological model consistent with all observations necessarily requires a description of gravity beyond Newton's theory. However, a cosmological Newtonian model, although it does not describe correctly the expansion of the universe, provides a first intuitive approach useful to appreciate the difference with the correct physical interpretation of the expansion of the universe within general relativity that we will discuss in Chapter 6,

3.1 NEWTON'S STATIC UNIVERSE

The first attempt to build a cosmological model within Newton's theory of gravity was proposed by Newton himself in the *Principia* [1] and today is referred to as Newton's static model.

In this model the universe is described as a uniform distribution of stars extending to infinity. The infinite size implies space homogeneity and isotropy, a necessary condition to have a net cancellation of all gravitational forces acting on each star, as required by a static universe. In this way the cosmological principle can be regarded as a consequence of the static condition. Therefore, time homogeneity is necessarily combined with space homogeneity and isotropy, obeying what is often referred to as the *perfect cosmological principle*.

However, such a model encounters various difficulties. The first is that because of Gauss' law, equivalent to inverse-square law of gravity, it is easy to show that the gravitational potential energy would be infinite at each point in the universe. Newton's static universe can be regarded as the limit of a spherical distribution of matter with radius R centred on a point where a generic star with mass $m_\star$ is located. The gravitational potential energy in this point would be given by

$$V(R) = -\frac{G\,M(R)\,m_\star}{R}.$$ (3.1)

Clearly, since $M(R) \propto R^3$, one has $V(R) \to -\infty$ for $R \to \infty$. This would also imply that an infinite force would act on the star if this is infinitesimally displaced from its equilibrium position. Therefore, the model cannot explain why the observed universe actually is far from homogeneity on scales below the size of superclusters of galaxies. This is not the only unphysical consequence of the Newtonian static universe.

Another very well-known one is *Olbers' paradox*. It basically consists of the contradiction between the assumption of a uniform, infinite distribution of stars and the sky being dark at night. Qualitatively, this can be stated by saying that, independently of the direction of observation, our sight-line should encounter a star eventually. One could reply that if a star is sufficiently far, then it appears so faint that the sky is bright only along those directions striking neighbouring stars. However, this counter-argument is clearly flawed, since it is easy to derive that observing the sky on a certain finite solid angle portion, for an infinite uniform distribution of stars the detected flux is constant. This is a straightforward consequence of the inverse-square law for the radiative flux.[1] Indicating with $n_\star$ the average number density of stars and with $L_\star$ their mean luminosity, the radiative flux $\Delta\Phi(r)$ of a shell of stars of radius r and thickness $\Delta r \ll r$, will be given approximately by

$$\Delta\Phi(r) \simeq \frac{L_\star}{4\pi r^2} \, 4\pi \, n_\star \, r^2 \, \Delta r = L_\star \, n_\star \, \Delta r \, . \tag{3.2}$$

In our homogeneous model, the *luminous intensity* of the shell is related to the radiative flux by $\Delta J(r) = \Delta\Phi(r)/(4\pi)$. Passing to the continuous limit ($\Delta r \to dr$) and integrating from $r = 0$ to $r = \infty$, one can immediately see that the total luminous intensity of any point of the sky would be infinite, again a clear unphysical result.

3.2 THE MILNE–McCREA MODEL

In an ultimate attempt to explain the expansion of the universe without having to abandon Newton's theory of gravity and at the same time avoiding the divergences appearing in the Newtonian static model, Milne and McCrea proposed in 1934 a model where the universe is made of a uniform gas, isotropically distributed around us within a certain finite radius R_U and everywhere uniformly expanding [2].

In the Milne–McCrea model, the cosmological principle is not respected since there is a centre of the universe clearly identifying a privileged position. This feature,

---

[1]Notice throughout the book that we adopt the definition of radiative or particle flux, introduced by Maxwell in kinetic theory. This is given by the current density vector $\vec{\Phi}$ associated to radiation or particles flow. The dimensionality is energy or number of particles per unit time per unit area. In the case of radiative flux, the current density is given by the Poynting vector. In the case of an astronomical source, its magnitude is given by luminosity per unit area. In the case of particle flux, one has $\vec{\Phi} = n\,\vec{v}$, where n is the number density and $\vec{v}$ is the velocity. Its magnitude is given by the number of particles crossing a surface, for example of a detector, per unit time per unit area. This definition should not be confused with the definition introduced in vectorial analysis and electromagnetism, where flux is the surface integral of the scalar product of the current density with the infinitesimal vector area and giving basically the charge. So-called luminous flux respects this second definition and corresponds to the same luminosity of the source.

necessary for the consistency of the model, should have been superseded in some later deeper version of the model.[2]

3.2.1 Everywhere uniform expansion

An important issue arises: is it possible to arrange an everywhere uniform expansion of a gaseous sphere?

Let us first consider a point-like test astronomical object (e.g., a galaxy) with mass m_i at a distance $r_i \leq R_U$ from the centre (in whose neighbourhood we are supposed to live). The maximum distance R_U can be regarded as the radius of the universe. The mass m_i will feel the gravitational attraction only from the matter contained inside the sphere of radius r_i around the centre with a mass

$$M(r_i) = \frac{4\pi}{3}\, \rho\, r_i^3 , \tag{3.3}$$

where ρ is the *mass density* of the gas of galaxies. This is assumed to be constant inside the sphere, at least on scales larger than the average intergalactic distance (homogeneous gas). The total energy $E_i(r_i)$ of the galaxy at a distance r_i from the centre will be therefore simply given by the sum of its kinetic energy and gravitational potential energy,

$$E_i = \frac{1}{2}\, m_i\, \dot{r}_i{}^2 - \frac{G\, M(r_i)\, m_i}{r_i} . \tag{3.4}$$

The condition that the expansion is everywhere uniform means that, independently of the galaxy and of its distance r_i, all galaxy positions scale in the same way. This implies that, if at some time $t_\star$ the galaxy is located at a distance $r_i(t_\star)$, then the ratio

$$a(t) \equiv \frac{r_i(t)}{r_i(t_\star)} = \frac{R_U(t)}{R_U(t_\star)} \quad \forall\, i , \tag{3.5}$$

is the same for all galaxies. In particular, it also applies to the same radius of the universe, as indicated. The function $a(t)$ is then a universal *scale factor* that describes the evolution of any galactic distance independently of its mass m_i. The value $r_{i\star} \equiv r_i(t_\star)$ is the *comoving coordinate* of the galaxy and remains constant during the expansion.

We can conveniently choose $t_\star = t_0$, our present time, in a way that $a(t) = r_i(t)/r_{i0} = R_U(t)/R_{U,0}$, implying $a_0 = 1$.

Why are we imposing the everywhere uniform expansion condition? Because it

---

[2]Ultimately, such a deeper model can be found only within general relativity. In this case a Newtonian description does not apply to the entire universe but only to spherical over or under-dense regions centred around any point of space and filled with an isotropic distribution of matter. This Newtonian limit can be rigorously justified under well precise conditions (Byrkhoff theorem) and provides a first simple description of the evolution of inhomogeneities on causally connected scales. Therefore, general relativity is anyway unavoidable in order to achieve a consistent cosmological model of the universe obeying the cosmological principle.

is quite straightforward to see that in this way Hubble's law follows immediately. We can indeed differentiate Eq. (3.5) obtaining

$$\dot{r}_i = v_i = \dot{a}\, r_{i0} = \frac{\dot{a}}{a}\, r_i \,. \tag{3.6}$$

If we now introduce the *expansion rate* $H(t) \equiv \dot{a}/a$, we can identify its present value with the Hubble constant, in a way that $H(t_0) = H_0$. By considering the non-relativistic Doppler effect relation $z = v/c$, we easily reproduce Hubble's law. This seems to be an encouraging result for the Milne–McCrea model.

If we now plug the scaling condition into the equation of energy conservation, and using Eq. (3.3), this becomes

$$E_i = \frac{1}{2}\, m_i\, r_{i0}^2\, \dot{a}^2 - \frac{4\,\pi\, G}{3}\, \rho\, m_i\, r_{i0}^2\, a^2 \,, \tag{3.7}$$

that can be rearranged in a way to obtain

$$H^2 - \frac{8\,\pi\, G}{3}\, \rho = \frac{C_i}{a^2} \,, \tag{3.8}$$

where $C_i \equiv 2\, E_i/(m_i\, r_{i0}^2)$. Notice that the left-hand side is independent of galactic positions and masses. Consistently, the right-hand side and in particular the C_i's, must be equal to a common constant that we can indicate with C_0. One can always introduce a *dimensionless* constant k and a constant length R_0 such that C_0 can be written as $C_0 \equiv -k\, c^2/R_0^2$.

Moreover, the constant length R_0 can always be chosen in a way that, if $k \neq 0$, then $|k| = 1$. With this convention, there are only three possible values for k: $k = +1$, $k = -1$, $k = 0$. The Eq. (3.8) can be finally written as

$$H^2 - \frac{8\,\pi\, G}{3}\, \rho = -\frac{k\, c^2}{a^2\, R_0^2} \,. \tag{3.9}$$

In this way this differential equation governing the time dependence of the scale factor looks formally equivalent to the *Friedmann equation* for a pressureless gas (usually called *dust* or *matter*) that we will derive within general relativity. However, the physical meaning of the equation and of the different quantities is very different.

The three possible values for k correspond to three different classes of models. From the relation Eq. (3.3), assuming that the total mass in the universe $M_U = M(R_U)$ is conserved and therefore constant, it immediately follows that

$$\rho(a) = \frac{\rho_0}{a^3} \propto a^{-3} \,. \tag{3.10}$$

In this way, Eq. (3.9) can be rewritten as

$$H^2 = \frac{8\,\pi\, G\, \rho_0}{3\, a^3} - \frac{k\, c^2}{a^2\, R_0^2} \,. \tag{3.11}$$

One can immediately see that the value of k determines the fate of the universe:

- If $k = -1$, then, starting at the present time with an expansion as observed ($H_0 > 0$), the universe will continue to expand forever;

- If $k = +1$, then there is a time where the scale factor reaches a maximum value given by

$$a_{\max} = \frac{8\pi\,G\,\rho_0\,R_0^2}{3\,c^2} = \frac{2\,G\,M_U\,R_0^2}{c^2\,R_{U,0}^3}\,.$$

- The borderline case $k = 0$ corresponds to a universe that expands forever, as for $k = -1$. However, in this case the asymptotic value of the expansion velocity $\dot{a}$ (i.e., its value for $t \to \infty$) vanishes.

Notice that if we differentiate Eq. (3.11), we easily recover, as expected, Newton's law

$$\ddot{a} = -\frac{4\pi\,G\,\rho_0}{3\,a^2} = -\frac{4\pi\,G\,\rho\,a}{3} < 0\,. \tag{3.12}$$

This confirms that the universe expansion in the Milne–McCrea model can only decelerate, a consequence of the fact that gravity is always attractive in Newton's theory. Another important observation is that in the Milne–McCrea model, cosmological redshifts are interpreted as a genuine Doppler effect.

3.2.2 CMB in the Milne–McCrea model

The Milne–McCrea model seems to be able to explain the cosmological expansion within a simple Newtonian description of gravity. However, we have to examine whether it can also explain the CMB.

Suppose that we interpret the CMB as a radiation that achieved thermal equilibrium (black body radiation) thanks to the existence of some kind of efficient process of emission and absorption within intergalactic dust. Notice that the energy conservation equation (3.7) for the energy of each galaxy, also implies a conservation equation for the total energy of the expanding universe.[3] In Eq. (3.7) we wrote the kinetic energy in the non-relativistic case and we did not include the mass energy term that is constant. If we do include the mass energy term as well, we can indicate the total energy of the matter in galaxies with U_M. The Eq. (3.7) implies then

$$\frac{dU_M}{dt} = 0\,. \tag{3.13}$$

The energy density $\varepsilon_M \equiv U_M/V_U$ of matter in galaxies will be dominated by the mass term and therefore will be given approximately by

$$\varepsilon_M \simeq \rho\,c^2\,, \tag{3.14}$$

where we neglected the kinetic energy term, corresponding to assume $p_M = 0$. This

---

[3]Assuming that the internal energy of each galaxy is negligible or in any case not exchanged during the expansion.

is because the velocities of stars and galaxies are well below the speed of light, reaching at most values of $\sim 1000 \, \text{km s}^{-1}$. In this way, from Eq. (3.10), one has $U_M \propto \varepsilon_M \, a^3$. Therefore, we can recast Eq. (3.13) as

$$\frac{d(\varepsilon_M \, a^3)}{dt} = 0 \quad \Leftrightarrow \quad \dot{\varepsilon}_M = -3 \, \varepsilon_M \, \frac{\dot{a}}{a}, \tag{3.15}$$

the *fluid equation for matter*.

How can this equation be generalised to account for thermal radiation as well? As we said, in order to explain the observed thermal equilibrium of CMB, we have to assume the existence of an efficient process coupling radiation with matter in galaxies during the expansion. This coupling would also explain that the universe is somehow opaque to radiation, that can in this way remain (approximately) trapped inside the sphere of radius R_U with negligible dissipation.

The expansion has to be quasi-static for thermal equilibrium to hold at all times. In this way we can apply the fundamental law of thermodynamics, writing

$$\frac{dU}{dt} = T \frac{dS}{dt} - p \frac{dV}{dt}, \tag{3.16}$$

where $U = U_M + U_\gamma$ is the total energy and $p = p_M + p_\gamma = p_\gamma$ is the total pressure. Considering that $dU_M/dt = 0$, Eq. (3.16) can also be written as

$$\frac{dU_\gamma}{dt} = T \frac{dS}{dt} - p_\gamma \frac{dV}{dt}. \tag{3.17}$$

Furthermore, if we assume that the expansion is adiabatic, implying that there are no heat exchanges between the expanding gas and the exterior, the total entropy is conserved and we can write (first law of thermodynamics)

$$\frac{dU_\gamma}{dt} = -p_\gamma \frac{dV}{dt}. \tag{3.18}$$

This implies that the gas of photons makes work during the expansion, losing energy. Introducing the photon energy density $\varepsilon_\gamma = U_\gamma/V$, and considering that $V = (4\pi/3) \, a^3(t) \, R_U(t_0)^3$, we can then write

$$\frac{d(\varepsilon_\gamma \, a^3)}{dt} = -p_\gamma \frac{da^3}{dt}. \tag{3.19}$$

From this one we can straightforwardly deduce the fluid equation for radiation

$$\dot{\varepsilon}_\gamma \frac{a}{\dot{a}} = -3 \, (p_\gamma + \varepsilon_\gamma). \tag{3.20}$$

Notice that the equation of state for radiation is $p_\gamma = \varepsilon_\gamma/3$, so that from the fluid equation one obtains

$$\varepsilon_\gamma \propto a^{-4}. \tag{3.21}$$

This differs from matter, for which $\varepsilon_M \simeq \rho \, c^2 \propto a^{-3}$. Since for radiation in thermal equilibrium $\varepsilon_\gamma \propto T^4$ (see Eq. (2.36)), Eq. (3.21) also implies $T \propto a^{-1}$.

If we introduce the total energy density $\varepsilon = \varepsilon_\gamma + \varepsilon_M$, the two fluid equations for matter and for radiation can be combined together yielding a general fluid equation[4]

$$\dot{\varepsilon} = -3\,(p + \varepsilon)\,\frac{\dot{a}}{a}\,. \tag{3.22}$$

Apparently, we managed to obtain a consistent picture of the expansion, also able to explain, potentially, the existence of CMB with a thermal equilibrium distribution.

3.2.3 Limitations of the Milne–McCrea model

At this point one can have the impression that within a Newtonian description of gravity, though with the objections listed in the beginning and though CMB existence is *potential* rather than *proven*, one can anyway obtain a reasonable understanding of the cosmological observations and Hubble's law.

The Milne–McCrea model seems to be a potential good candidate for a cosmological model based on Newtonian gravity. However, it is certainly incomplete since in any case one should also be able to include relativistic effects. These are certainly unavoidable in the description of the electromagnetic field, but there are also much simpler reasons originating from the model itself. Since galactic velocities cannot be higher than the speed of light, the size of the universe is necessarily upper bounded by

$$R_{\mathrm{U}}(t) < R_H(t) \equiv c\,H^{-1}(t)\,, \tag{3.23}$$

where $R_H(t)$ is the *Hubble radius* that at present is given by $R_{H,0} = c\,H_0^{-1} \simeq 4\,\mathrm{Gpc}$. However, we have seen that, in the case $k = -1$, the expansion would last forever and asymptotically $H(t \to \infty) > 0$. This necessarily implies that at some time the upper bound Eq. (3.23) would be violated. Therefore, a relativistic extension of the model is unavoidable. One could ignore the problem barring the case $k = -1$ but there is an additional issue calling for a relativistic extension. Although using the simple non-relativistic Doppler effect relation, $z = v/c$, we could reproduce Hubble's law, experimentally we do observe objects with redshifts $z \gtrsim 1$. These of course cannot be described by the non-relativistic Doppler effect, since it would imply unphysical, superluminal galactic velocities. Therefore, the observed high redshifts force us to introduce relativistic effects. In the case of Doppler effect, one knows how this gets generalised to the relativistic case. One simply has

$$z = \sqrt{\frac{1 + \beta}{1 - \beta}} - 1\,, \tag{3.24}$$

where $\beta \equiv v/c$. In this way, for $\beta \to 1$, one obtains arbitrarily large redshifts. Hubble's law would then become

$$z = \sqrt{\frac{1 + H_0\,d/c}{1 - H_0\,d/c}} - 1\,. \tag{3.25}$$

---

[4]It could also be obtained directly from Eq. (3.16).

This is a possibility that can be tested with observations of astronomical objects with high redshifts, such as SNIa. However, we will see in Chapter 10 that even including relativistic effects, a Doppler effect interpretation still fails in correctly reproducing current data. Even neglecting this observational test, there is still a big conceptual issue. We started from a non-relativistic description of gravity, but relativistic effects are present anyway. The model, therefore, necessarily requires a relativistic extension of Newton's theory of gravity and this extension is general relativity. As we will discuss in the next chapter, in general relativity, when photons propagate in the presence of a gravitational field, they experience a *gravitational redshift*, something intrinsically different from Doppler effect. This description of cosmological redshifts correctly reproduces the observed cosmological redshifts, for any value, as we will discuss in Chapter 10.

In addition to the problem of cosmological redshifts, there are further objections (some we mentioned already) to the Milne–McCrea model that we summarise here:

- It abandons the cosmological principle reintroducing a centre of the universe;

- Since it relies on non-relativistic Newton's theory of gravity, it cannot consistently account for a relativistic motion of particles;

- Hubble's law is reproduced rather than explained, since it seems to come out accidentally, as a result of a fine-tuned choice of initial conditions for the motion of different galaxies;

- The same objection applies to the existence of CMB, relying on an unjustified matter-radiation coupling that can only be achieved, as we will discuss, at temperatures much higher than T_0.

- It does not address the problem of primordial nuclear abundances.

All these objections are strong reasons not to persist with the Milne–McCrea model and, more generally, with the idea of a cosmological model based on Newton's theory of gravity. However, the Milne–McCrea model can be regarded as a preparatory, Newtonian analogy of Friedmann cosmological models based on general relativity. More interestingly, the equations we derived, after proper reinterpretation, can also be derived within general relativity but in the description of the evolution of perturbations rather than of the entire universe.

In particular we will again derive the fluid equation, but the interpretation of the scale factor and expansion will be physically very different. The Friedmann equation will get generalised replacing the mass density ρ with the total energy density ε. Apparently, this seems a minimal modification, but actually it has a striking consequence. It leads to an acceleration equation intrinsically different from Newton's law (see Eq. (3.12)), where the acceleration of the expansion depends on a contribution to the energy density from radiation (light gravitates!) and even from pressure (pressure gravitates as well!). This is a clear sign that the physical interpretation of cosmological expansion within a relativistic description, requires a profound conceptual modification compared to a Newtonian interpretation, not just a minimal one.

On the other, as mentioned already, within a cosmological model based on general relativity, it can be rigorously shown (Byrkhoff's theorem) why the Milne–McCrea model manages to grasp some basic features of the cosmological expansion. These are, however, applicable rigorously not to the universe as a whole, but to local density perturbations. This also motivates the discussion we did of the Milne–McCrea model. In conclusion, it will gradually become clear, during our discussion, how cosmology represents a (further) phenomenological triumph for general relativity.

EXERCISES

Exercise 3.1 *Suppose you are in an infinitely large static universe in which the number density of stars is $n = 10^9 \, \text{Mpc}^{-3}$ and the average stellar radius is equal to the Sun's radius $R_\odot = 3 \times 10^8 \, \text{m}$. How far, on average, could you see in any direction before your sight-line struck a star?*

Exercise 3.2 *What would be the size of the Universe in the Milne–McCrea model with $H_0 = 70 \, \text{km s}^{-1} \, \text{Mpc}^{-1}$ and highest observed redshift $z = 0.1$?*

Exercise 3.3 *A rough estimate of the mass density in the Universe today could be based on the typical galaxy mass and galaxy separation $\rho_0 = M_{\text{gal}}/\text{Mpc}^3$. Evaluate this expression in eV m^{-3}.*

Exercise 3.4 *If $H_0 = 70 \, \text{km s}^{-1} \, \text{Mpc}^{-1}$, and given the density found in the previous problem, what would be the ultimate fate of the Universe in a Milne–McCrea model, would it expand forever or reach a maximum size and then start a contraction phase?*

Exercise 3.5 *Show that in the non-relativistic limit, for $\beta \to 0$, Eq. (3.24) correctly reproduces the non-relativistic limit $z = \beta$.*

BIBLIOGRAPHY

[1] Newton, Isaac, *Philosophiae Naturalis Principia Mathematica (Mathematical Principles of Natural Philosophy)*, London, 1687; Cambridge, 1713; London, 1726. (Pirated versions of the 1713 edition were also published in Amsterdam in 1714 and 1723.)

[2] E.A. Milne, *A Newtonian Expanding Universe*, Quart. J. Math. Oxford **5**, 64-72 (1934); W.H. McCrea and E.A. Milne, *Newtonian Universes and the Curvature of Space*, Quart. J. Math. Oxford **5**, 73-80 (1934).

[3] A. Einstein, *Relativity: The special and the general theory, a popular exposition*, authorised translation by R.W. Lawson, Methuen & Co. Ltd., London

(1920), Part III: *Consideration on the Universe as a Whole*, Chapter XXX. *Cosmological difficulties of Newton's theory.*

CHAPTER 4

From classical mechanics to relativistic theories

In this chapter we start discussing some theoretical aspects of classical mechanics, such as the definition of vectors and tensors. We then show how they get generalised in special relativity. In this way, in the last section we will be able discuss in a self-contained way basic ingredients of general relativity, necessary for a proper comprehension and derivation of Friedmann cosmological models. This combined discussion should highlight the genuine novel features of general relativity, making it possible to understand common features of the three theories and how they realise a sort of set of Chinese boxes.

4.1 SCALARS, VECTORS AND TENSORS IN CLASSICAL MECHANICS

Let us shortly review how scalars, vectors and tensors are defined in classical mechanics [1].

4.1.1 The Galilean space-time

Classical mechanics is based on the assumption that physical systems can be described within the *Galilean space-time*. Given a reference frame $\mathcal{S}$, an event is specified by a time coordinate t and by a position vector $\vec{x}$.

It is always possible to find *orthonormal* reference frames such that the infinitesimal distance dr between two points can be simply written as[1]

$$dr^2 = dx_1^2 + dx_2^2 + dx_3^2 = \sum_{i,j} \delta_{ij} \, dx_i \, dx_j \,, \tag{4.1}$$

(where δ_{ij} is the Kronecker Delta function), expressing the so called *Euclidean metric*.

---

[1]We are following the general convention, adopted throughout the book, of indicating space components with Latin indexes running from 1 to 3.

4.1.1.1 Symmetries of the Galilean space-time

The Galilean space-time exhibits some important symmetries.

- *Space is homogeneous.* This means that if an experiment is first performed in a certain region of space and then the same is repeated translating all (really all!) components by the same distance (i.e., in a rigid way), the results, before and after the translation, are exactly the same. This is also equivalent to saying that there are no privileged points in space.

- *Space is isotropic.* This means that if an experiment is first performed with a certain space orientation of all components and then the same is repeated upon a rigid rotation of all of them, the results are identical. This is equivalent to saying that there are no privileged space orientations.

  Homogeneity and isotropy of space imply the existence of an infinite number of *inertial reference frames*, where a body placed at rest in a certain point remains at rest forever.[2]

- *Time is homogeneous.* This means that the results of two identical experiments do not change if these differ just by a time shift. In order for the two experiments to be identical, it is crucial that all their experimental components obey the same initial conditions at the two different initial times. [3]

- *Principle of relativity.* This means that if an experiment is first performed and then repeated with all experimental components rigidly boosted by a uniform linear velocity, one would obtain identical experimental results. This is also equivalent to saying that laws of physics are invariant under a boost transformation.[4]

4.1.1.2 Galilean transformations

Consider a generic transformation from a reference frame S with origin O to a reference frame S' with origin O', obtained combining a rotation described by a

---

[2]Suppose that a body initially at rest would start moving, in a given point, in a certain direction with a certain velocity. Because of homogeneity there must be a transformation to a reference frame where such a motion has to be the same for all points. This direction would identify a privileged direction in disagreement with isotropy. For this reason it has always to be possible to make a transformation to a system where a body initially at rest remains at rest forever (*inertial reference frame*) and from this one can obtain an infinite set of them differing by origin position and axis orientation.

[3]Assuming that this postulate is correct, it means that if two seemingly identical experiments performed in different times give different results, then necessarily there was some uncontrolled variation of the external conditions. This provides an experimental strategy to reveal the presence of unknown and unwanted sources of background that can undergo, for example, some seasonal variation: it is not just a formal aspect!

[4]This fundamental postulate has been beautifully depicted by Galileo in his *Dialogue* [2] in terms of an observer in a ship hold (with closed windows) who would not be able to realise whether the ship is at rest in the harbour or sailing with a perfectly uniform linear motion in an ideally calm sea.

rotation matrix R, a (rigid) translation described by a displacement $\vec{d}$ of the origin, and finally a uniform motion (boost) with velocity $\vec{V}$. A generic event, with space-time coordinates $(t, \vec{x})$ in $\mathcal{S}$, will have space-time coordinates $(t', \vec{x}')$ in $\mathcal{S}'$ given by the *general Galilean transformations*[5]

$$t' = t, \tag{4.2}$$
$$\vec{x}' = R\vec{x} - \vec{d} - \vec{V}t. \tag{4.3}$$

The first equation is a consequence of the fact that in classical mechanics the interactions can propagate at infinite velocity. Therefore, independently of the boost velocity $\vec{V}$, two simultaneous events in $\mathcal{S}$ will be simultaneous also in $\mathcal{S}'$ and vice versa. This is the reason why there is a universal time t ($t' = t$). This can be equivalently expressed saying that time is invariant under Galilean transformations, or also that it is a *(non-relativistic) scalar invariant* or simply a *scalar*.

4.1.1.3 Scalars and vectors in classical mechanics

There are many other important scalars in classical mechanics. Given two events, $(t_A, \vec{r}_A)$ and $(t_B, \vec{r}_B)$, the space distance $d_{AB}^{(3)}$ between them, defined as the length of the distance vector $\vec{r}_A - \vec{r}_B$, namely

$$d_{AB}^{(3)} \equiv \sqrt{\sum_i (x_{Ai} - x_{Bi})^2} = \sqrt{\sum_i (x'_{Ai} - x'_{Bi})^2} = d_{AB}^{(3)'}, \tag{4.4}$$

is also a scalar under Galilean transformations. In particular, the length of the position vector is invariant under spatial rotations[6] $\vec{x}' = R\vec{x}$. Therefore, imposing $|\vec{x}'|^2 = |\vec{x}|^2$, it follows that R must be an *orthogonal matrix*, i.e., it must respect the orthogonality condition $R^T R = I$ or equivalently, in components, $\sum_k R_{ki} R_{kj} = \delta_{ij}$.

A generic vector $\vec{a}$ is defined as a set of three quantities that obey the same transformation law as $\vec{x}$ under rotations, i.e., $\vec{a}' = R\vec{a}$. It is easy to verify that the scalar product of two vectors is a scalar and in particular the length $|\vec{a}|$ of a vector is a (non-relativistic) scalar.

An important example of scalar quantity is the *mass m* of an elementary particle. Indicating its position function with $\vec{x}(t)$ and considering that time is a scalar quantity, its *velocity* $\vec{v} \equiv d\vec{x}/dt$ is a vector. *Momentum*, defined as $\vec{p} = m\vec{v}$, is also an example of vector quantity.

Consider now an *element of fluid*, i.e., a portion of a fluid containing a sufficiently large number of elementary particles that the thermodynamic limit can be

---

[5] We are barring a time translation with the origin of time shifted by a time interval τ.
[6] Notice that in components the rotation can be written as

$$x'_i = \sum_j R_{ij} x_j, \tag{4.5}$$

with the convention, adopted throughout the book, that Latin indexes run from 1 to 3 in sums.

considered valid but at the same time of sufficiently small size that the spatial variation of thermodynamic quantities is negligible. One can then introduce its position $\vec{x}(t)$, velocity $\vec{v}(\vec{x}, t)$ and describe its thermodynamic state in terms of scalar quantities like mass density $\rho(\vec{x}, t)$ and pressure $p(\vec{x}, t)$, and also in terms of vectorial quantities like $\vec{j} \equiv \rho\,\vec{v}$, called *mass flux density*. From *conservation of mass* one can easily derive the *continuity equation*

$$\frac{\partial \rho}{\partial t} = -\vec{\nabla}(\rho\,\vec{v})\,, \tag{4.6}$$

simply stating that the variation of mass in a small volume centred on $\vec{x}$, must be given by the net flux, the difference between the mass flowing in and the mass flowing out. The continuity equation is therefore a consequence of mass conservation and of the fact that mass flows continuously; it does not jump from one point to another. This reminder of *fluid mechanics* provides not only an example of the use of scalars and vectors in classical mechanics, but it will also be useful for our cosmological applications.

4.1.1.4 Tensors in classical mechanics

A *tensor* T_{ij} is a physical object that can be represented by a 3×3 matrix whose components transform under spatial rotations like[7]

$$T'_{ij} = \sum_{l,m} R_{il}\,R_{jm}\,T_{lm}\,. \tag{4.7}$$

If T_{ij} is a tensor then, given a vector $\vec{a}$, the quantities

$$b_i = \sum_j T_{ij}\,a_j \tag{4.8}$$

must form a vector and this provides an easy way to test whether, given a (3×3) matrix, this is a tensor or not. An example of a tensor, that can be simply built from two generic vectors $\vec{a}$ and $\vec{b}$, is the so-called *direct product*, whose components are given by

$$T_{ij} = a_i\,b_j\,, \tag{4.9}$$

or in a compact matrix notation $T_{ij} = (a\,b^T)_{ij}$, where a is a column vector and b^T is a row vector.[8] Another example of a tensor is the *cross product* of two vectors. For example, given two vectors $\vec{a}$ and $\vec{b}$, the cross product is defined as

$$(\vec{a} \times \vec{b})_{ij} = a_i\,b_j - a_j\,b_i\,. \tag{4.10}$$

---

[7]More precisely this is a *rank 2 tensor* where the rank of a tensor is the number of indices that are needed in order to label tensor components. From this point of view scalars can be regarded as *rank 0 tensors* and vectors as *rank 1 tensors*. However, when we generically refer to a tensor, we will always imply a rank 2 tensor.

[8]An equally valid alternative definition is $T_{ij} = (b\,a^T)_{ij}$.

Notice that the cross product is an *antisymmetric tensor* since $(\vec{a} \times \vec{b})_{ij} = -(\vec{a} \times \vec{b})_{ji}$ and in particular the diagonal components vanish, i.e., $(\vec{a} \times \vec{b})_{ii} = 0$. In this way, it has only three independent components that form what is called an *axial vector*. Therefore, an axial vector (or pseudo-vector) is rigorously speaking an antisymmetric tensor and not a true genuine vector: it is a sort of tensor disguised as a vector.

A particularly important tensor is the *stress tensor* σ. It extends the definition of *pressure* in fluid dynamics, taking into account the *shear forces* acting on an element of fluid. The entries of σ give the components of the force $\vec{F}_{\hat{n}}$ that acts on the surface of a fluid element with a unit normal vector $\hat{n}$,

$$F_{\hat{n}i} = \sum_j \sigma_{ij} \, n_j \, . \tag{4.11}$$

It is a *symmetric tensor* (i.e., $\sigma_{ij} = \sigma_{ji}$). In the case of vanishing *viscosity*, the fluid is called *ideal fluid* and in this case the stress tensor is simply given by

$$\sigma_{ij} = p \, \delta_{ij} \, , \tag{4.12}$$

where p is pressure. In this case there are no shear forces, force is always perpendicular to the surface and the same on any surface element.

Another example of a tensor that is encountered in classical mechanics is the *tensor of inertia I*, giving the angular momentum $\vec{L}$ of a rigid body in terms of the angular velocity $\vec{\omega}$,

$$\vec{L} = I\vec{\omega}, \quad \text{or in components} \quad L_i = \sum_j I_{ij} \, \omega_j \, . \tag{4.13}$$

Notice that a tensor is represented by a (3×3) matrix in a specific reference frame but it is more than just a matrix, in the same way as a vector is more than just a column or row vector. In the next sections we will generalise the definition of tensors within special relativity and general relativity but it should be clear from this section that tensors are already present in classical mechanics though to a less fundamental level than in special relativity and general relativity. The reason is simply that the fundamental equations of classical mechanics are vectorial equations, while in special relativity Maxwell's equations become tensorial equations and also in general relativity the fundamental equations for the gravitational field, Einstein's equations, are tensorial equations. This is why one can have the (wrong!) impression that tensors are specific to relativistic theories but it is obviously not true. Tensors are simply unavoidable in relativistic theories while they are avoidable (and sometimes avoided just to simplify things) in classical mechanics.

4.2   A REMINDER OF SPECIAL RELATIVITY

Before discussing general relativity, it will prove useful to revise the basic postulates of special relativity [3]. We want moreover to emphasise some aspects of special relativity that will be afterwards particularly important within general relativity, like for example the definition of an ideal fluid in relativistic fluid dynamics and the role of the energy-momentum tensor. It will be easier in this way to understand the novelties introduced by general relativity that are particularly important in cosmology.

4.2.1 Postulates

The postulates of special relativity *almost* coincide with those of classical physics. An Euclidean geometry is still assumed to hold for the three-dimensional space that is still assumed to be homogeneous and isotropic. Time also is still assumed to be homogeneous.

The first aspect that we need to discuss in special relativity is how to choose a *reference frame*. First, imagine a grid of Cartesian coordinate lines in space with a watch measuring time at each coordinate point.

As discussed in the previous section, according to the Galilean *principle of relativity*, the laws of classical mechanics are invariant under a Galilean boost transformation. With the discovery of the laws of electromagnetism, beautifully described in a unified picture by Maxwell's equations, there was a common belief that these were violating principle of relativity. In particular, it was thought that the speed of electromagnetic waves (i.e., of light) had to be defined with respect to some medium, the so-called *aether*, filling all space. Therefore, the aether identifies a privileged reference frame setting an absolute origin of velocities, analogously to the case of the propagation of sound waves in the air. In this case, it was expected that changing the relative velocity of the observer with respect to the aether, would have resulted in a variation of the measured speed of light. However, the Michelson–Morley experiment found no evidence of such variation, with quite high precision. Einstein's main motivation was the formulation of a theory where the principle of relativity applies also to electromagnetism explaining the result of the Michelson–Morley experiment.

Notice that, as in classical mechanics, since the space is assumed to be homogeneous and isotropic, there must exist an *inertial reference frame* where a particle at rest remains at rest. Because of the assumption of homogeneity and isotropy combined with the principle of relativity, there must be an infinite number of such reference frames connected by transformations combining rotations, translations and boosts. The whole set of these transformations is called inhomogeneous Lorentz transformations or Poincarè transformations. In full generality they also include discrete transformations such as parity and time reversal; otherwise, if restricted to continuous transformations, they are called *proper* Poincarè transformations. These transformations must leave unchanged all laws of physics, including electromagnetism and are called *global symmetries of the space-time*.

Another important postulate, common to both classical mechanics and special relativity, is the *principle of causality*, the assumption that the interactions of physical systems are causal.

The crucial difference between special relativity and classical mechanics is that in the second case the speed of the causal message (the maximum speed of a propagating interaction) is infinite, while in the case of special relativity it is assumed to have a maximum value. From electromagnetism, the maximum value is identified with the *speed of light*, given by[9]

$$c = 299792458 \, \mathrm{m\,s^{-1}} \,. \tag{4.14}$$

Because of the principle of relativity, this maximum value has to be a universal quantity, the same in all inertial reference frames and the same independently of the velocity of the emitting source and observer.

We can then use light to synchronise all watches in any given inertial reference frame. The finiteness and universality of the speed of light has the straightforward consequence that time is not any more an absolute quantity as in classical mechanics, but it changes in a transformation from a reference frame to a boosted one.

4.2.2 Simultaneous events

This statement is easy to understand considering a simple thought experiment (*gedankenexperiment*) [4]. Suppose that you are on a train travelling with constant velocity v and that you are located just in the middle of a coach $2\ell_0$ long. Two light signals are *simultaneously* emitted from the front and from the back of the coach toward you and will reach you *simultaneously* after a time ℓ_0/c.

In that moment the train is crossing a crowded station with people standing with their watches. From the point of view of the outside observers the two signals travel exactly at the same speed. Therefore, they will reach simultaneously a point inside the coach, at rest with respect to the external observers, that was coinciding with the middle of the coach when signals were emitted but not when they arrive. Therefore, from their point of view the only possibility to explain that the two signals reach you at the same time is that they were not emitted simultaneously but with a time difference Δt that can be easily calculated.

The time interval measured by the external observers standing in the station for the signal emitted from the back of the train to reach you is given by

$$\Delta t_{\mathrm{back}} = t_{\mathrm{arrival}}^{\mathrm{back}} - t_{\mathrm{emission}}^{\mathrm{back}} = \frac{\ell_0}{c - v} \,, \tag{4.15}$$

and for the signal emitted from the front

$$\Delta t_{\mathrm{front}} = t_{\mathrm{arrival}}^{\mathrm{front}} - t_{\mathrm{emission}}^{\mathrm{front}} = \frac{\ell_0}{c + v} \,. \tag{4.16}$$

Therefore, the external observers conclude that they were emitted with a time difference

$$\Delta t = \Delta t_{\mathrm{back}} - \Delta t_{\mathrm{front}} = t_{\mathrm{emission}}^{\mathrm{front}} - t_{\mathrm{emission}}^{\mathrm{back}} = \frac{2\,\ell_0\,v}{c^2 - v^2} \neq 0 \,. \tag{4.17}$$

---

[9]It is an exact value, since the meter is defined as the distance travelled by light in 299792458^{-1} s.

In conclusion, *two events, that appear to be simultaneous in one reference frame, are not necessarily simultaneous in a boosted reference frame.*

Notice that in classical mechanics this does not happen. The two intervals of time, Δt_{back} and Δt_{front}, would either both vanish assuming an infinite speed of the causal message (so that also $\Delta t = 0$) or, assuming light to have finite speed, they would be equal. This is because the speed of the emitted signal measured by the external observers, would be the result of a Galilean composition of the velocity of light with the velocity of the train. Therefore, the front signal would have a speed $c - v$ instead of c and the back signal would appear to move with a velocity $c + v$. In this way one would obtain the same result for Δt_{front} and Δt_{back} measured by the external observers at rest in the station.

4.2.3 Minkowski space

Introducing an origin O in the $3+1$ space-time, we can associate a *4-position vector* x to any generic event. If we also introduce an orthonormal reference frame $\mathcal{S}$, we can further associate a *4-coordinate vector*

$$x^\mu \equiv (x^0, x^1, x^2, x^3) = (x^0, \vec{x})\,, \tag{4.18}$$

where $x^0 \equiv ct$, $\vec{x} = (x^1, x^2, x^3) = (x, y, z)$ and c is the *speed of light*. We can then define the *interval* between an event A and an event B as

$$d(A, B) \equiv s_{AB} \equiv \sqrt{(x_B^0 - x_A^0)^2 - (x_B^1 - x_A^1)^2 - (x_B^2 - x_A^2)^2 - (x_B^3 - x_A^3)^2}\,. \tag{4.19}$$

The interval can be regarded as a measure of the distance between two events in Minkowski space. The *Minkowski metric*, i.e., the infinitesimal interval element in Minkowski space, is given by

$$ds^2 = (dx^0)^2 - d\ell^2\,. \tag{4.20}$$

It is a so-called *pseudo-Euclidean* metric contrarily to the usual Euclidean metric of the three-dim subspace given by

$$d\ell^2 = (dx^1)^2 + (dx^2)^2 + (dx^3)^2 \tag{4.21}$$

that satisfies Pythagoras' theorem. Introducing the *Minkowski metric tensor*

$$\eta_{\alpha\beta} = \begin{pmatrix} 1 & 0 & 0 & 0 \\ 0 & -1 & 0 & 0 \\ 0 & 0 & -1 & 0 \\ 0 & 0 & 0 & -1 \end{pmatrix}\,, \tag{4.22}$$

the Minkowski metric Eq. (4.20) can be recast as

$$ds^2 = \eta_{\alpha\beta}\, dx^\alpha\, dx^\beta\,, \tag{4.23}$$

where we introduced the Einstein rule of implying the sum on saturated indexes (also called repeated or dummy indices).

Thesis: *The interval is an invariant under the symmetries of the space-time, the Poincarè transformations.* Let us prove this statement [4].

It is clearly invariant under space-time translations and spatial rotations affecting only the 3-spatial coordinates (i.e., $dt = dx^0 = 0$ in this case). Thus we have simply to prove that it is invariant under boost transformations. This can be done in two steps.

The first step is to notice that if an interval vanishes in one reference frame S then it must also vanish in a reference frame S' that moves with respect to S with a uniform velocity $\vec{V}$. This clearly follows from the universality of the speed of light.

The event A can be associated to the emission of a light signal from a point at some given time and the event B to its detection. If, for simplicity, we choose A to be the origin of S, then one clearly has

$$\sum_i (x_B^i)^2 = (x_B^0)^2 \tag{4.24}$$

and therefore $s_{OB}^2 = 0$. Since the speed of light is the same in all inertial reference frames, then in S' we can also analogously write

$$\sum_i (x_B'^i)^2 = (x_B'^0)^2, \tag{4.25}$$

so that $s_{OB}'^2 = 0$ as well.

The second step is slightly more involved. First of all let us notice that the result of the first step, that we just derived in general, can be specialised to the case of an infinitesimal interval, so that if $ds^2 = 0$ then $ds'^2 = 0$ and vice versa. This necessarily implies $ds'^2 = b\, ds^2$, in a way that if one vanishes the other vanishes as well. Notice that b cannot depend on the specific event x^μ, otherwise space-time homogeneity would be violated. Therefore, b can only depend on the boost velocity $\vec{V}$, so that we can write $b = b(\vec{V})$. However, imposing the isotropy of the space one can even more restrictively conclude that b can only depend on the magnitude of the boost velocity $V \equiv |\vec{V}|$, so that one has

$$ds'^2 = b(V)\, ds^2. \tag{4.26}$$

Finally, from the principle of relativity, one can say that in the inverse transformation $S' \to S$, the infinitesimal interval has to transform exactly in the same way but this time with boost velocity $-\vec{V}$, so that we can write

$$ds^2 = b(-\vec{V})\, ds'^2 = b(V)\, ds'^2. \tag{4.27}$$

Finally, using the direct transformation Eq. (4.26), we arrive to

$$ds^2 = b^2(V)\, ds^2, \tag{4.28}$$

trivially implying $b(V) = 1$, i.e., $ds^2 = ds'^2$. The invariance of an infinitesimal interval then necessarily implies also the invariance of a finite interval. In this way we finally proved the thesis.

4.2.4   Lorentz boost along the x-axis

In special relativity a boost transformation is commonly referred to as *Lorentz boost.* Let us apply the invariance of the interval to a special Lorentz boost, $\mathcal{S} \to \mathcal{S}'$, where the boost velocity $\vec{V}$ is oriented along the x-axis, so that $\vec{V} = (V, 0, 0)$. In this transformation the variables y and z do not change. Therefore, the invariance of the interval implies

$$(x^0)^2 - x^2 = (x^{0'})^2 - x'^2 \,. \tag{4.29}$$

This is satisfied by the following transformation (equivalent to a rotation in the plane (it, x)),

$$
\begin{aligned}
x^{0'} &= x^0 \cosh \psi + x \sinh \psi \\
x' &= x^0 \sinh \psi + x \cosh \psi \,.
\end{aligned} \tag{4.30}
$$

The motion of the spatial origin O' is described by $x'_{O'} = 0$ and, therefore,

$$\tanh \psi = -\frac{x}{x^0} = -\frac{x}{ct} = -\frac{V}{c} \equiv -\beta \,, \tag{4.31}$$

from which it follows

$$\cosh \psi = \frac{1}{\sqrt{1 - \beta^2}} \equiv \gamma \quad \text{and} \quad \sinh \psi = -\frac{\beta}{\sqrt{1 - \beta^2}} \equiv -\beta \gamma \,, \tag{4.32}$$

where we introduced the *Lorentz factor* $\gamma \equiv 1/\sqrt{1 - \beta^2}$. Therefore, the Lorentz transformation for a boost along the x-axis is given by

$$
\begin{aligned}
x^{0'} &= \gamma \left(x^0 - \beta \, x \right), \\
x' &= \gamma \left(x - \beta \, x^0 \right),
\end{aligned} \tag{4.33}
$$

while clearly the corresponding inverse Lorentz transformation is given by

$$
\begin{aligned}
x^0 &= \gamma \left(x^{0'} + \beta \, x' \right), \\
x &= \gamma \left(x' + \beta \, x^{0'} \right).
\end{aligned} \tag{4.34}
$$

From these expressions one recovers the Galilean transformations in the limit $c \to \infty$. Moreover, one can see that necessarily $V < c$ and this is a check confirming that the transformations are correct, since we imposed from the beginning, as a first principle, that the maximum speed is c. Notice that $V = c$ is also not allowed, meaning that one cannot have an inertial reference frame moving at the speed of light with respect to the observer. This is a consequence of the fact that light travels at the speed of light in all reference frames and so one cannot choose a particular reference frame where light is at rest: photons cannot be stopped! The boost transformation along the x-axis can be of course easily generalised to a generic boost transformation.

4.2.5 Proper time

Consider a point-like particle moving in an inertial reference frame $\mathcal{S}$ with generic velocity $\vec{v}(t)$. We can always find at each time t a new inertial reference frame $\mathcal{S}'(t)$, where the particle is located in the origin and therefore $x'^i = 0$. If the particle moves with a uniform motion, then $\mathcal{S}'(t)$ coincides with the same inertial reference frame at all times and one recovers a simple Lorentz boost transformation. Otherwise, for arbitrary motion, it should be clear that at each time t, there is a different (instantaneous) inertial reference frame $\mathcal{S}'(t)$ with boost velocity with respect to $\mathcal{S}$ given by $\vec{V}(t) = \vec{v}(t)$ and Lorentz factor $\gamma(t) = 1/\sqrt{1 - |\vec{v}|^2(t)/c^2}$. Within an infinitesimal time dt, the particle travels a distance $d\ell = \sqrt{\sum_i (dx^i)^2}$ in $\mathcal{S}$, while in $\mathcal{S}'$, where it is at rest, one has $dx'^i = 0$.

Therefore, from the invariance of the interval, we can write

$$ds^2 = c^2\, dt^2 - d\ell^2 = c^2\, d\tau^2 \Rightarrow d\tau = \frac{ds}{c}\,, \tag{4.35}$$

where τ is the time that would be measured by a hypothetical watch moving jointly with the particle. This time is called *proper time*. Since $|\vec{v}(t)| = d\ell/dt$, from $c^2\, dt^2 - d\ell^2 = c^2\, d\tau^2$, one finds

$$d\tau = \frac{dt}{\gamma(t)} \leq dt\,, \tag{4.36}$$

showing that the proper time interval is always smaller than the time interval measured by observers at rest in $\mathcal{S}$, except for the trivial case $\vec{v}(t) = 0$, implying $\gamma(t) = 1$, when they are equal. This means that the hands of a watch moving jointly with the particle would be seen by observers at rest in $\mathcal{S}$, along the path covered by the particle, moving slower than the hands of their watches.

Suppose now that you set an experiment to measure a characteristic time period of a physical system, for example the life-time of a decaying particle. Suppose that this is first measured by an observer in a reference frame $\mathcal{S}'$ where the system is at rest, for example in the origin, and that the result is τ. Suppose now that the measurement is repeated by observers at rest in a reference frame $\mathcal{S}$, where the system is seen to move with constant velocity $\vec{v}$. From Eq. (4.36), one immediately obtains that the corresponding measured time period *in flight* is given by

$$t = \gamma \tau > \tau\,. \tag{4.37}$$

This effect is called *dilation time*.[10] In particular, the life time of an elementary particle measured in flight is always longer than the life time at rest. Moreover notice that while τ is a universal invariant property of the system, t depends on the velocity of the system in $\mathcal{S}$.

Consider now the special case when a point-like particle starts moving in a reference frame $\mathcal{S}$ at time t_{start}, it travels along a closed path with generic velocity

---

[10]This result can also be easily derived from the Lorentz transformations.

$\vec{v}(t)$ and, finally, it is back to the starting point at time t_{arrival}. At the end of the journey, the elapsed proper time interval is given by

$$\Delta \tau = \int_{t_{\text{start}}}^{t_{\text{arrival}}} \frac{dt}{\gamma(t)} < \Delta t \equiv t_{\text{arrival}} - t_{\text{start}}. \tag{4.38}$$

This result implies the so-called twin paradox: an astronaut who returns home after a space trip at a speed close to the speed of light, will be much younger than his twin who remained on the Earth. This is a paradox because one could expect, according to the principle of relativity, an equivalence, since both twins move with respect to each other. However, notice that there is no equivalence between the two twins: one is always at rest in an inertial reference frame while the other is necessarily moving with non uniform speed. In conclusion, flying on a fast plane makes you younger.

4.2.5.1 Relativistic Doppler effect

As an application of the dilation time effect, it is useful to derive Eq. (3.24) describing the redshift of a moving luminous object, the so-called *relativistic Doppler effect*. The emitted wave length of the radiation can be written as $\lambda_{\text{em}} = cT$, where T is the oscillation period of the e.m. field. If the object is receding, after one wavefront is emitted (e.g., when the electric field is maximal along a transversal direction), it will have moved a distance vT before the next wavefront is emitted. This is a first effect to be taken into account and that is non-vanishing even in the non-relativistic limit giving rise to the non-relativistic Doppler effect. In special relativity one has also to take into account the dilation effect. The period of oscillation of the e.m. field that we indicate with T', measured in the reference frame where the observer is at rest and where the luminous object is moving with a recession velocity v, is different from T and given by $T' = \gamma T$. Taking into account both two effects, one obtains

$$\lambda_{\text{obs}} = cT' + vT' = \gamma cT(1 + \beta) = \lambda_{\text{em}} \sqrt{\frac{1 + \beta}{1 - \beta}}. \tag{4.39}$$

From this expression, Eq. (3.24) easily follows. One can see how the non-relativistic limit is recovered from the relativistic case in the limit $\beta \ll 1$. In this limit the dilation time effect is negligible since it is described by γ and is therefore a $\mathcal{O}(\beta^2)$ effect. Therefore, one is just left with the first effect that does not vanish even in the non-relativistic limit, since it is a $\mathcal{O}(\beta)$ effect.

4.2.6 Proper length

Analogously to the proper time, one can also define a *proper length*. For example, consider the length $\ell_0 \equiv \ell(\vec{V} = 0)$ of a ruler measured at rest in the inertial reference frame $\mathcal{S}$ chosen in such a way that the ruler lies along the x-axis and its two edges have coordinates $x = x_1$ and $x = x_2$, so that $\ell_0 = \Delta x = x_2 - x_1$.

Let us consider now what is the length of the ruler measured in $\mathcal{S}'$, an inertial reference frame that moves with velocity $\vec{V} = (V, 0, 0)$ with respect to $\mathcal{S}$. The quantity $\ell(\vec{V}) \equiv x_2' - x_1'$ is the length of the ruler measured by two observers at rest in $\mathcal{S}'$, therefore moving with respect to the ruler with a speed $\vec{V}$, and located at the two edges at the same time t'. If we use the inverse Lorentz transformation, we can write

$$x_2 = \gamma(x_2' + \beta\,c\,t'), \tag{4.40}$$

$$x_1 = \gamma(x_1' + \beta\,c\,t'), \tag{4.41}$$

and, simply by subtracting the two equations, we find

$$\Delta x = \ell_0 = \gamma\,\ell \Rightarrow \ell(V) = \frac{\ell_0}{\gamma(V)} < \ell_0. \tag{4.42}$$

This shows that a length measured in flight is always shorter than proper length: this effect is called *Lorentz contraction*.[11]

4.2.7 General Lorentz transformations

The Lorentz transformations along the x-axis (cf. Eqs. (4.33) and (4.34)) can also be written in a compact matrix form as

$$x'^{\mu} = \Lambda^{\mu}{}_{\nu}\,x^{\nu}, \tag{4.43}$$

where the matrix Λ is given by

$$\Lambda \equiv \Lambda^{\mu}{}_{\nu} = \begin{pmatrix} \gamma & -\gamma\,\beta & 0 & 0 \\ -\gamma\,\beta & \gamma & 0 & 0 \\ 0 & 0 & 1 & 0 \\ 0 & 0 & 0 & 1 \end{pmatrix}. \tag{4.44}$$

The adopted notation to indicate the Λ matrix elements with one index 'up' and one 'down' will be explained soon. For the time being it can be regarded just as a notational definition.

From the invariance of the interval, it can be shown (Exercise 4.8) that the most general Lorentz transformation, including spatial rotations, can still be written like

$$x'^{\mu} = \Lambda^{\mu}{}_{\nu}\,x^{\nu}, \tag{4.45}$$

---

[11] A simple particle physics application of time dilation and Lorentz contraction is given by atmospheric muon decays. The muon lifetime at rest is $\tau_{\mu} \simeq 2.2 \times 10^{-6}$s. They are produced in the outer layers of the atmosphere at least 100 km above ground level at relativistic speeds. Without time dilation, these particles should travel only distances lower than $c\,\tau_{\mu} \simeq 600$ m before decaying and so they would be unable to reach the ground. However, ground detectors do detect them! The puzzle is easily solved considering that in our reference frame their lifetime is dilated to $\gamma\,\tau_{\mu}$. Since some of them are produced to energies such that their Lorentz factor $\gamma \gtrsim 200$, these muons can indeed survive reaching the ground and be detected before decaying. In a reference frame where the muon is at rest, the lifetime has clearly the rest value but the atmosphere thickness is contracted by a factor γ and the final physical result is the same as before, in agreement with the principle of relativity.

where the matrix Λ, with entries Λ^μ_ν, has to satisfy the condition

$$\Lambda^T \eta \Lambda = \eta, \tag{4.46}$$

where η is the Minkowski metric tensor. In this relation a matrix product is implied. It can also be written explicitly as

$$\Lambda^\nu_\alpha \eta_{\nu\mu} \Lambda^\mu_\beta = \eta_{\alpha\beta}, \tag{4.47}$$

where remember we are using Einstein's convention for repeated indexes (in this case μ and ν).[12]

If one includes space-time translations, then the most general transformation can be written as

$$x'^\mu = \Lambda^\mu_\nu x^\nu + a^\mu, \tag{4.49}$$

where a^μ is a simple displacement 4-vector. These transformations are called Poincarè transformations.

Notice that in general these transformations also include *discrete transformations*. These are transformations that cannot be obtained from the identity applying a sequence of infinitesimal transformations. Discrete transformations are parity, time reversal, reflections. The difference between continuous Lorentz transformations and discrete Lorentz transformations is that in the first case $\det(\Lambda) = 1$ and in the second case $\det(\Lambda) = -1$.

If one neglects discrete transformations, one obtains so-called proper Lorentz transformations (when translations are excluded) and proper Poincarè transformations (when space-time translations are included).

4.2.8 Scalars and 4-vectors (in special relativity)

Analogously to what we have seen in classical mechanics, one can define (Lorentz) scalars, 4-vectors and 4-tensors, depending on how they transform under Lorentz transformations.

A *Lorentz scalar* is a quantity that is invariant under Lorentz transformations. We have already seen an example: the interval s^2_{AB} between two events A and B is by definition a (Lorentz) scalar like the metric ds^2.

A *contravariant 4-vector* A^μ is a set of four components that under Lorentz transformations transform like the components of the 4-coordinate vector, i.e.,

$$A^\mu \to A'^\mu = \Lambda^\mu_\nu A^\nu. \tag{4.50}$$

---

[12]Notice that the condition

$$\Lambda^T \eta \Lambda = \eta, \tag{4.48}$$

where η is the Minkowski metric tensor, is the relativistic generalization of the orthogonality condition $R^T R = I$ holding for a rotation matrix R. In this respect, a Lorentz transformation Λ has to be regarded as a kind of rotation in the Minkowski space-time. This analogy becomes complete if one considers a space-time where the usual real time is replaced by the complex time it, in this case the metric tensor η is replaced by the identity (in 4-dimensions), i.e., a Euclidean metric tensor in 4 dimensions. In many contexts this equivalence is used as a useful trick in order to simplify calculations.

Therefore, by definition, the 4-position coordinate vector is a 4-vector. The space components of a contravariant 4-vector transform as a 3-vector $\vec{A}$, so that one can write $A^\mu = (A^0, \vec{A})$.

From any contravariant 4-vector, whose components are denoted with an upper index, we can also define a *covariant 4-vector*

$$A_\mu \equiv \eta_{\mu\nu} A^\nu \,, \tag{4.51}$$

whose components are denoted with a lower index. Notice that if $A^\mu = (A^0, \vec{A})$, then $A_\mu = (A^0, -\vec{A})$, i.e., the time component of a contravariant 4-vector is equal to the time component of the corresponding covariant 4-vector but the space components change sign.

Hence contravariant components A^μ and covariant components A_μ are two different ways to represent the same 4-vector A.[13] From two generic 4-vectors A and B, one can build a scalar called *scalar product*, defined as

$$A \cdot B \equiv \eta_{\mu\nu} A^\mu B^\nu \,. \tag{4.52}$$

Using covariant components, the scalar product of two 4-vectors can also be written, in a more compact way, as $A \cdot B = A^\mu B_\mu = A_\mu B^\mu$.

In particular, the *squared length* of a 4-vector A, defined as

$$A^2 \equiv A \cdot A = A^\mu A_\mu = (A^0)^2 - |\vec{A}|^2 \,, \tag{4.53}$$

is a scalar. For example the quantity $x_A \cdot x_A = x_{A\mu} x_A^\mu$, i.e., the interval between the origin and the event A, is a scalar.

An important 4-vector is the *4-velocity*, defined as

$$u^\mu \equiv \frac{dx^\mu}{ds} \,, \tag{4.54}$$

that is related to the usual 3-velocity simply by[14]

$$u^\mu = (\gamma, \gamma \vec{\beta}) \,, \tag{4.57}$$

where we defined $\vec{\beta} \equiv \vec{v}/c$ (in the natural system simply $\vec{\beta} = \vec{v}$). Notice that, with this definition, the 4-*velocity* is dimensionless. It is easy to prove that

$$u^2 = 1 \,, \tag{4.58}$$

---

[13]It is said that between covariant and contravariant components there is a *duality* and the 4-vector A is a *dual* vector.

[14]Using Eq. (4.35) and Eq. (4.36), the time component is given by

$$u^0 \equiv \frac{dx^0}{ds} = \frac{dt}{d\tau} = \gamma \,, \tag{4.55}$$

while the space components are given by

$$\vec{u} \equiv \frac{d\vec{x}}{ds} = \frac{1}{c}\frac{d\vec{x}}{d\tau} = \frac{\gamma}{c}\frac{d\vec{x}}{dt} = \gamma \frac{\vec{v}}{c} = \gamma \vec{\beta}. \tag{4.56}$$

that is obviously an invariant.

One can also define the 4-*acceleration*

$$a^\mu \equiv \frac{du^\mu}{ds}.$$ (4.59)

For a free particle one has simply $a^\mu = 0$, generalising the result valid in classical mechanics that a free point-like particle moves uniformly.

Another important 4-vector is the 4-*momentum* p^μ of a point-like particle of mass m, defined as

$$p^\mu \equiv m\,c\,u^\mu = (\gamma\,m\,c, \gamma\,m\,\vec{v}).$$ (4.60)

Its squared length is given by $p^2 = m^2\,c^2$. In the non-relativistic limit the zero component gives the non-relativistic energy of the particle (divided by c). Therefore, one can identify the zero component of the 4−momentum with the relativistic energy E of a free particle, while the spatial components can be identified with the relativistic 3-momentum $\vec{p}$, so that[15]

$$E = \gamma\,m\,c^2, \qquad \vec{p} = \gamma\,m\,\vec{v}.$$ (4.61)

Therefore, the expression for squared length of the 4-momentum, $p^2 = m^2\,c^2$ is equivalent to the well-known relation

$$E^2 = c^2\,|\vec{p}|^2 + m^2\,c^4.$$ (4.62)

The kinetic energy is simply defined as $K \equiv E - m\,c^2$. In the non-relativistic limit, for $|\vec{p}|\,c \ll m\,c^2$, one obtains

$$E \simeq E_{nr} \equiv m\,c^2 + \frac{|\vec{p}|^2}{2\,m},$$ (4.63)

so that simply $K = |\vec{p}|^2/(2m)$. Conversely in the ultra-relativistic limit, for $m\,c^2 \ll |\vec{p}|\,c$, one obtains

$$E \simeq E_{ur} \equiv |\vec{p}|\,c + \frac{m^2\,c^4}{2\,|\vec{p}|\,c}.$$ (4.64)

In particular, for massless particles like photons,[16] one has exactly $E = c\,|\vec{p}|$.

In *natural units* these relations greatly simplify, becoming

$$p^\mu = (E, \vec{p}) = (\gamma\,m, \gamma\,m\,\vec{\beta})$$ (4.65)

and $E^2 - |\vec{p}|^2 = m^2$. In particular for massive particles at rest ($|\vec{p}| = 0$) one simply has $E = m$, while for massless particles like photons one has $E = |\vec{p}|$.

---

[15]Notice that we are indicating with p the length of the 4-momentum and with $|\vec{p}|$ the length of the 3-momentum.

[16]Basically photons are the only particles that we know to be certainly massless. The lightest neutrino could be also massless since current experiments just place an upper bound given by $m_{\nu_i}\,c^2 \lesssim 0.1\,\text{eV}$ on each neutrino mass (there are three neutrino masses that we can indicate with $m_{\nu_1} \leq m_{\nu_2} \leq m_{\nu_3}$). On the other hand, we know that the other two neutrino masses are certainly non-vanishing, since current neutrino oscillation experiments place a lower bound given by $m_{\nu_2} \gtrsim 0.008\,\text{eV}/c^2$ for the next-to-lightest neutrino and $m_{\nu_3} \gtrsim 0.05\,\text{eV}/c^2$ for the heaviest one.

4.2.9 Tensors

Tensors generalise the definition of vectors. A quantity $T^{\mu\nu}$ is a *rank-2 tensor with 2 contravariant indexes* if in a Lorentz transformation it transforms like

$$T'^{\mu\nu} = \Lambda^{\mu}_{\ \alpha} \Lambda^{\nu}_{\ \beta} T^{\alpha\beta}. \tag{4.66}$$

Indexes can be lowered as for 4-vectors and for example,

$$T_{\mu\nu} = \eta_{\mu\alpha} \eta_{\nu\beta} T^{\alpha\beta} \tag{4.67}$$

obtaining a rank-2 tensor with 2 covariant indexes. Notice that we could have also lowered only one index as well, obtaining a mixed tensor with one contravariant (up) index and one covariant (low). As before, one can obtain 4-vectors or scalars from tensors *contracting* indexes. For example, from a tensor $T^{\mu\nu}$ we can obtain a scalar $\mathcal{T}$, the *trace* of the tensor T, contracting the two indexes, explicitly

$$\mathcal{T} \equiv \eta_{\mu\nu} T^{\mu\nu} = T^{\mu}_{\ \mu}. \tag{4.68}$$

4.2.10 The energy-momentum tensor

As we have seen, energy and momentum of a single point-like particle is described in terms of its 4-momentum vector $p^{\mu} = (E/c, \vec{p})$. If we have a distribution of point-like particles with momenta $p^{\mu}_{(n)}$ (here the bracketed subscript n labels the nth-particle), the so-called *energy-momentum* tensor, at a certain space-time point x, is defined as

$$T^{\mu\nu}(x) = \sum_n \frac{p^{\mu}_{(n)}(t) \, p^{\nu}_{(n)}(t)}{E_{(n)}(t)} \, \delta^3(\vec{x} - \vec{x}_{(n)}(t)). \tag{4.69}$$

For a fluid, a continuous distribution of matter-radiation, the energy-momentum tensor is the relativistic generalization of the stress tensor in classical mechanics and is given by[17]

$$T^{\mu}_{\ \nu} = \begin{pmatrix} \varepsilon & S_x/c & S_y/c & S_z/c \\ S_x/c & -\sigma_{xx} & -\sigma_{xy} & -\sigma_{xz} \\ S_y/c & -\sigma_{yx} & -\sigma_{yy} & -\sigma_{yz} \\ S_z/c & -\sigma_{zx} & -\sigma_{zy} & -\sigma_{zz} \end{pmatrix}, \tag{4.70}$$

where ε is the energy density, $\vec{S}$ is the energy flux vector and σ_{ij} are the components of the stress tensor. This looks quite a complicated expression but fortunately we are interested only in the case of an *ideal fluid* for which

$$T^{\mu\nu} = -p\,\eta^{\mu\nu} + (p + \varepsilon)\, u^{\mu}\, u^{\nu}, \tag{4.71}$$

---

[17]We imply the dependence on x in this and next expressions.

where u^μ is the 4-velocity of the element of fluid in the given point and p is pressure already introduced in classical mechanics. In the reference frame where the element of fluid is at rest, the expression Eq. (4.71) for $T^{\mu\nu}$ gets simplified into

$$T^{\mu\nu} = \begin{pmatrix} \varepsilon & 0 & 0 & 0 \\ 0 & p & 0 & 0 \\ 0 & 0 & p & 0 \\ 0 & 0 & 0 & p \end{pmatrix}. \tag{4.72}$$

If we lower one index, it simply becomes

$$T^\mu{}_\nu = \begin{pmatrix} \varepsilon & 0 & 0 & 0 \\ 0 & -p & 0 & 0 \\ 0 & 0 & -p & 0 \\ 0 & 0 & 0 & -p \end{pmatrix}. \tag{4.73}$$

In the non-relativistic limit one has $p \ll \varepsilon$ and $\varepsilon \simeq \rho c^2$, so that

$$T^\mu{}_\nu \simeq \begin{pmatrix} \rho c^2 & 0 & 0 & 0 \\ 0 & 0 & 0 & 0 \\ 0 & 0 & 0 & 0 \\ 0 & 0 & 0 & 0 \end{pmatrix}, \tag{4.74}$$

where remember that ρ is the mass density.

The relativistic extension of the equation of continuity (4.6) is given by the *energy-momentum conservation equation*.

$$\frac{\partial T^{\mu\nu}}{\partial x^\nu} \equiv T^{\mu\nu}{}_{,\nu} = 0. \tag{4.75}$$

This yields a *fluid equation* that relates p, ε and their partial derivatives.

It can be shown that it is basically a consequence of the first principle of thermodynamics,

$$d(\varepsilon V) = -p\, dV, \tag{4.76}$$

applied to an element of fluid with sufficiently small volume V that the fluid inside can be considered homogeneous. The differential dV accounts for the fact that the fluid can undergo compression or dilatation if there is no hydrostatic equilibrium. In the non-relativistic limit one has $p \to 0$, $\varepsilon \simeq \rho c^2$ and Eq. (4.76) reduces to a mass conservation equation that, consistently, implies the continuity equation.

4.2.11 Relativistic action and equations of motion for a free particle

The action S of a system is defined as

$$S = \int_A^B ds\, L, \tag{4.77}$$

where L is the Lagrangian of the system and the integral is calculated on paths in space-time, called *world-lines*, between two events A and B.

The equations of motion are obtained from the *principle of least action* saying that the variables of motion of the system will evolve following the world-line that minimises the action (the *physical* world-line), therefore imposing $\delta S = 0$. From the principle of least action one can derive the equations of motion in terms of the Lagrangian (the Lagrange equations).

Let us consider the simplest system: what is the relativistic action of a free (point-like) particle? Since we want our equations to satisfy the principle of relativity, the action has to be necessarily the integral of a scalar. The only scalar we can write for the Lagrangian of a free particle is

$$L = \alpha\, \eta_{\alpha\beta}\, u^\alpha\, u^\beta. \tag{4.78}$$

Since we want to re-obtain in the non-relativistic limit the usual Lagrangian of a free particle, $L = mv^2/2$, one must have $\alpha = -m\,c^2$.

The Lagrangian equations are then given by

$$\frac{d}{ds}\frac{\partial L}{\partial u^\mu} = \frac{\partial L}{\partial x^\mu}. \tag{4.79}$$

The right-hand side vanishes since the Lagrangian cannot depend on the position (it is a consequence homogeneity!) and, therefore, one simply obtains in a formal way that the relativistic motion of a free particle has to be of the form

$$a^\mu \equiv \frac{du^\mu}{ds} \equiv \frac{d^2 x^\mu}{ds^2} = 0, \tag{4.80}$$

where, as we have seen, a^μ is the 4-acceleration.

Later on we will see some simple examples of relativistic particle kinematics applied to the early universe.

4.2.12 Maxwell's equations in a covariant form

The laws of electromagnetism, encoded by Maxwell's equations, can also be formulated in a covariant way, this means in a way to respect the principle of relativity and that the speed of light is manifestly independent of the relative motion between the source and the observer, as found experimentally by Michelson and Morley.

Introducing the electromagnetic field

$$F_{\mu\nu} = \begin{pmatrix} 0 & E_1/c & E_2/c & E_3/c \\ -E_1/c & 0 & -B_3 & B_2 \\ -E_2/c & B_3 & 0 & -B_1 \\ -E_3/c & -B_2 & B_1 & 0 \end{pmatrix}, \tag{4.81}$$

inhomogeneous Maxwell's equations (in the SI) in a covariant form are given by

$$\partial_\mu F^{\mu\nu} = \mu_0\, J^\nu, \tag{4.82}$$

and homogeneous Maxwell's equations by

$$\partial^\lambda F^{\mu\nu} + \partial^\mu F^{\nu\lambda} + \partial^\nu F^{\lambda\mu} = 0 \,, \tag{4.83}$$

where $J^\mu \equiv (c\,\rho_{\text{el}}, \vec{J})$ is the 4-current density vector, with ρ_{el} the electric charge density and $\vec{J}$ the electric 3-current density. The electric field $\vec{E} \equiv (E_1, E_2, E_3)$ and the magnetic field $\vec{B} \equiv (B_1, B_2, B_3)$ appear, in a unified fashion, as different components of the electromagnetic field $F_{\mu\nu}$, that is a (relativistic) tensor. In this way they mix in the Lorentz transformations, analogously to how time and space 4-position vector components mix. For example a charge at rest will produce an electric field described by the Coulomb law and in a boosted reference frame where the electric charge is in motion, this will also generate a magnetic field.

In this way Einstein managed to fully accomplish his initial objective of obtaining a more general theory where the principle of relativity applies to electromagnetism as well, but a new even harder challenge was waiting for him.

4.3 ELEMENTS OF GENERAL RELATIVITY

After achieving a covariant formulation of classical mechanics and electromagnetism, Einstein turned to the investigation of a covariant formulation that would also include gravity, providing a relativistic generalisation of Newton's law. This was certainly a first motivation behind the theory of general relativity.

There was, however, a second motivation that is stated quite clearly at the beginning of Einstein's paper on *The Foundation of the General Theory of Relativity* in 1916 [5]. This was the idea of a *generally covariant theory*, a theory where the fundamental equations are covariant for a general coordinate transformation, not only between inertial reference frames. In this way Einstein wanted to try to address the problem of the existence of privileged inertial reference frames,[18] where the equations need to be written in order to be Lorentz invariant. Quoting Einstein's words: *general laws of nature are to be expressed by equations which hold good for all systems of coordinates, that is, are covariant with respect to any subsitution, i.e., general coordinate transformation, whatever.*[19] We will explain in the following what should be meant by general coordinate transformation.

The two motivations are closely related to each other and this is mainly a consequence of the principle of equivalence.

---

[18]The existence in classical mechanics of inertial reference frames sets an absolute origin of acceleration. The idea that such origin could be removed by a more general theory was at the basis of the *Mach's principle*, a conjecture suggesting that inertial forces originate from the cosmological distribution of matter around us. Mach's principle certainly provided an inspiration to Einstein's formulation of general relativity, in fact it is explicitly discussed in Section 2 of [5], but it should not be regarded as a fundamental principle of general relativity, rather as an initial guiding idea.

[19]This does not exclude that their solution can be much simpler within a suitable reference frame.

4.3.1 Coordinate transformations in the absence of a gravitational field

We have seen that in special relativity, using Cartesian coordinates, the metric can be written as

$$ds^2 = \eta_{\mu\nu} \, dx^\mu \, dx^\nu \,. \tag{4.84}$$

We have also seen that Lorentz transformations leave the metric invariant.

Suppose now you perform an arbitrary coordinate transformation $x^\mu \to x'^\mu(x^\nu)$ such that the inverse transformation exists and is given by

$$x^\mu = x^\mu(x'^\nu) \,. \tag{4.85}$$

In this new set of general coordinates the metric would now read as

$$ds^2 = \eta_{\mu\nu} \frac{\partial x^\mu}{\partial x'^\alpha} \frac{\partial x^\nu}{\partial x'^\beta} \, dx'^\alpha \, dx'^\beta \,. \tag{4.86}$$

This can also be written as

$$ds^2 = g_{\alpha\beta} \, dx'^\alpha \, dx'^\beta \,, \tag{4.87}$$

where we introduced

$$g_{\alpha\beta} \equiv \eta_{\mu\nu} \frac{\partial x^\mu}{\partial x'^\alpha} \frac{\partial x^\nu}{\partial x'^\beta} \,, \tag{4.88}$$

the *metric tensor* in the new coordinate system. Therefore, for an arbitrary coordinate system, the Minkowski metric tensor is replaced by this new metric tensor.

Let us discuss a very simple example considering a transformation involving only spatial coordinates, converting between Cartesian coordinates and spherical polar coordinates (see Fig. 4.1).

Explicitly, this means that one performs the transformation $(t, x, y, z) \to (t, r, \theta, \phi)$, where θ and ϕ are respectively polar and azimuthal angle. The Cartesian coordinated x, y, z can be expressed in terms of r, θ, ϕ as

$$
\begin{aligned}
x &= r \sin\theta \cos\phi \,, \\
y &= r \sin\theta \sin\phi \,, \\
z &= r \cos\theta \,.
\end{aligned}
\tag{4.89}
$$

Looking at Fig. 4.1, it is straightforward from Pythagoras' theorem to write[20]

$$d\ell^2 = dr^2 + r^2 \, d\theta^2 + r^2 \, \sin^2\theta \, d\phi^2 \,. \tag{4.90}$$

The metric will then read in the new coordinates as

$$ds^2 = c^2 \, dt^2 - (dr^2 + r^2 \, [d\theta^2 + \sin^2\theta \, d\phi^2]) \,. \tag{4.91}$$

---

[20] Alternatively, one can check directly that the same result is found differentiating the transformations Eqs. (4.89).

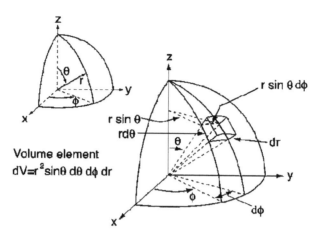

FIGURE 4.1: Spherical polar coordinates (from *Hyperphysics*, C.R. Nave).

The metric tensor $g_{\mu\nu}$ in the new set of coordinates is then given by

$$
g_{\mu\nu} = \begin{pmatrix}
1 & 0 & 0 & 0 \\
0 & -1 & 0 & 0 \\
0 & 0 & -r^2 & 0 \\
0 & 0 & 0 & -r^2 \sin^2\theta
\end{pmatrix}.
\tag{4.92}
$$

Of course such a transformation does not pretend to describe any new physical process. It could be just simply more convenient to be adopted for the description of some particular problem (e.g., a satellite orbiting around the Earth). Therefore, so far, we cannot see any real physical advantage or opportunity in performing a change of coordinates as general as Eq. (4.85).

4.3.2 Equivalence principle

The important point made by Einstein is that such general transformations could be used to describe how physics laws change in the presence of a gravitational field. In this way he could arrive at a completely new description of gravity where the gravitational interaction is entirely described in terms of the space-time metric tensor, however, not of the type we have seen before, where we simply converted between Cartesian and a new set of coordinates.

The first important step is the *equivalence principle*. This can be stated at different levels. At the lowest one it was already noticed by Galileo, Kepler and Newton. For clarity, it is useful to make a comparison between electrostatic and gravitational forces.

Let us first write Coulomb's law for the electrostatic force acting on a point-like electric charge q_{el} due to the electric field generated by a second electric charge Q_{el}

in the origin,[21]

$$\vec{f}_{el} = \alpha_{el} \frac{Q_{el} \, q_{el}}{r^2} \, \hat{r} \, . \tag{4.93}$$

Newton's law for the gravitational force acting on a massive body with *gravitational charge* q_{gr} and due to the gravitational field generated by a point-like particle with gravitational charge Q_{gr} can be written in a very similar way

$$\vec{f}_{gr} = -\alpha_{gr} \frac{Q_{gr} \, q_{gr}}{r^2} \, \hat{r} \, . \tag{4.94}$$

However, there are two very important differences:

1) The electric charge can be either positive or negative and, therefore, the force can be either attractive or repulsive while in the gravitational case the gravitational charge can only be positive and, for this reason, the gravitational force is always attractive;

2) If one considers Newton's second law of motion for the acceleration of the gravitational charge q_{gr} with *inertial mass* m, one can write

$$\vec{a} = \frac{\vec{f}_{gr}}{m} \, , \tag{4.95}$$

where $\vec{f}_{gr}$ is given by Eq. (4.94) in the case of a point-like gravitational source. It is found experimentally that, independently of the material and of the value of m, the acceleration $\vec{a}$ is the same for all bodies, implying

$$\frac{q_{gr}}{m} = \text{const} \, . \tag{4.96}$$

One can then always rescale α_{gr} in a way to have exactly $q_{gr} = m$ and in this case α_{gr} can be identified with Newton's constant G. For this reason one usually refers indistinctly either to the gravitational charge or to the inertial mass simply as *mass*.

Einstein started from this important experimental observation, tested today with huge accuracy, to go one step further. Consider a particle at rest in an isolated box (you can think for example of Galileo's ship as a box). If there is an equivalence between the gravitational charge and the inertial mass of a body, this means that one cannot determine whether the box is uniformly moving in an inertial reference frame or free falling in a homogeneous gravitational field (one should think for example of the gravitational field close to the Earth's surface removing completely air friction since it has to be a free fall), this is the *equivalence principle in a weak formulation*.

One could object that the Earth's gravitational field is not exactly homogeneous

---

[21]In the SI one has $\alpha_{el} = (4\pi\epsilon_0)^{-1}$. However, this is an irrelevant point here, you should focus on a comparison between the two laws.

and therefore, because of tidal forces, one should observe a small motion of the bodies in the box towards the boundary walls. This is indeed true and, for this reason, one has to specify that the box has to be small enough with respect to the typical length scale of the gravitational field variation so that the (weak) equivalence principle holds in the limit of an infinitesimal box size.

Notice moreover that the (weak) equivalence principle also implies equivalently that one cannot detect any difference between a particle in a box on the Earth's surface feeling a gravitational attraction and a particle in the same box (imagine a rocket) accelerating upward with an acceleration equal to the same gravity acceleration. Therefore, more formally, the (weak) equivalence principle can also be rephrased saying that *for a (sufficiently small) free falling body in a gravitational field there is an exact cancellation between the gravitational and the inertial force.* Let us translate this in equations. In an inertial frame Newton's law reads

$$m \frac{d^2 \vec{x}}{dt^2} = q_{\text{gr}} \, \vec{g} \,, \tag{4.97}$$

and, since $m = q_{\text{gr}}$, we can always make the coordinate transformation

$$\tilde{\vec{x}} = \vec{x} - \frac{1}{2} \, \vec{g} \, t^2 \tag{4.98}$$

such that in the new (free falling) coordinate system, a non-inertial frame, one has

$$\frac{d^2 \tilde{\vec{x}}}{dt^2} = 0 \,, \tag{4.99}$$

recovering exactly the same law of motion valid in an inertial reference frame in the absence of gravitational field. This illustrates explicitly the equivalence principle in the weak version.

Einstein even made a step forward and formulated the *equivalence principle in a strong formulation* (what is usually meant by equivalence principle) stating that *given an arbitrary gravitational field it is always possible to find in any point of the space-time a so-called locally inertial reference frame (also called free-falling system) where, in a sufficiently small region around the point, the laws of physics are the same as in an inertial reference frame in the absence of gravitation (i.e., those found in special relativity).* From this principle, one can deduce a perfectly consistent relativistic theory of gravity, as we are going to discuss.

4.3.3 Equations of motion for a point-like particle in a gravitational field: the geodesic equation

We have seen that Eq. (4.80) gives the equation of motion for a free particle in special relativity. In the presence of a gravitational field, this can be generalised imposing the equivalence principle. Hence, there must be a *locally inertial coordinate system*, denoted with ξ^μ, where the equation of motion can be written in the same form as Eq. (4.80) for a free particle, explicitly

$$\frac{d^2\xi^\mu}{ds^2} = 0 \,. \tag{4.100}$$

Also the metric has to reduce to the Minkowski metric in ξ^μ coordinates,

$$ds^2 = \eta_{\alpha\beta}\, d\xi^\alpha\, d\xi^\beta \,. \tag{4.101}$$

We can say that the free-falling system generalises the global transformation Eq. (4.98) to a non-inertial reference frame with uniform acceleration that, thanks to the equivalence principle, cancels the gravitational field. However, such a cancellation works exactly only in a given point, since a real gravitational field is not constant in space. You can think for example of the Earth's gravitational field

$$\vec{g}(\vec{r}) = -\frac{M_\oplus G}{r^2}\, \hat{r} \,. \tag{4.102}$$

For a real gravitational field, one has to choose, for any point, a different non-inertial coordinate system so that $\xi^\mu = \xi^\mu(x^\nu)$. In this new locally inertial system[22] all properties of special relativity have to be recovered.

Converting between locally inertial coordinates ξ^μ and general coordinates x^μ, the Minkowski metric transforms into

$$ds^2 = g_{\mu\nu}\, dx^\mu\, dx^\nu \,, \tag{4.103}$$

with the metric tensor $g_{\mu\nu}$ given by

$$g_{\mu\nu} = \eta_{\alpha\beta}\, \frac{\partial\xi^\alpha}{\partial x^\mu}\, \frac{\partial\xi^\beta}{\partial x^\nu} \,. \tag{4.104}$$

In these coordinates one can show that the equations of motion Eq. (4.100) become

$$\frac{d^2\, x^\mu}{ds^2} = -\Gamma^\mu_{\alpha\beta}\, \frac{dx^\alpha}{ds}\, \frac{dx^\beta}{ds} \,, \tag{4.105}$$

where $\Gamma^\mu_{\alpha\beta}$ are the *affine connections* (or *Christoffel symbols*), defined as

$$\Gamma^\mu_{\alpha\beta} \equiv \frac{\partial\xi^\mu}{\partial x^\nu}\, \frac{\partial^2\xi^\nu}{\partial x^\alpha\, \partial x^\beta} \,. \tag{4.106}$$

These are symmetric under the exchange of α with β because of the Schwarz integrability condition and are related to the metric tensor by the expression[23]

$$\Gamma^\mu_{\alpha\beta} = \frac{1}{2}\, g^{\mu\sigma}\, (g_{\sigma\alpha,\beta} + g_{\beta\sigma,\alpha} - g_{\alpha\beta,\sigma}) \,. \tag{4.107}$$

---

[22]You can visualise for example thinking of watches located in some of the coordinate points that are free falling onto the Earth from all directions.

[23]We are using the common compact notation $X(x^\nu)_{,\alpha} \equiv \partial X/\partial x^\alpha$ for the derivatives of a generic quantity $X(x^\nu)$ with respect to space-time coordinates.

The equation (4.105) is also called the *geodesic equation* (the reason will be clear later on). It is the relativistic extension of Newton's second law of motion and one expects that in some limit it reproduces it. The gravitational interaction is completely encoded in the *metric tensor* $g_{\mu\nu}$, that can be therefore identified with the *gravitational field*, the analogue of the electromagnetic field $F_{\mu\nu}$ in electromagnetism. In the case of gravity, however, the gravitational field, identified with the metric tensor as a consequence of the equivalence principle, plays at the same time another important role: it also describes the geometry of the space-time.

4.3.4 The Newtonian limit

If we write the gravitational acceleration $\vec{g} \equiv \vec{f}_{\text{gr}}/m$ in terms of the Newtonian gravitational potential field ϕ, so that $\vec{g} = -\vec{\nabla}\phi$, Newton's law Eq. (4.95) can be written as

$$\frac{d^2\vec{x}}{dt^2} = -\vec{\nabla}\phi. \tag{4.108}$$

The geodesic equation (4.105) is expected to be a generalisation of Newton's law for the acceleration of a point-like particle with mass m subject to a gravitational force. Therefore, one expects that in some limit, the geodesic equation can reproduce Newton's law. Let us show that this *Newtonian limit* can be recovered under the following three conditions:

- Non-relativistic motion;

- Weak gravitational field;

- Stationary gravitational field (the metric tensor is not rapidly changing with time).

The non-relativistic limit implies that in the right-hand side of the geodesic equation we can neglect all the terms containing spatial velocities components dx^i/ds. In this way, only the (time-time) term $\alpha = \beta = 0$ is non-vanishing in Eq. (4.105) and

$$\frac{d^2 x^\mu}{ds^2} \simeq -\Gamma^\mu_{00} \left(\frac{dx^0}{ds}\right)^2. \tag{4.109}$$

The second condition requires the gravitational field to be *weak*. How can that be expressed? If the gravitational field is switched off, then the metric has to reduce to the Minkowski metric, since we have to recover the same laws of special relativity. Therefore, for a weak gravitational field, the metric tensor has to differ from the Minkowski metric tensor just by a small perturbation $h_{\mu\nu}$, and we can write

$$g_{\mu\nu} = \eta_{\mu\nu} + h_{\mu\nu} \quad \text{with } |h_{\mu\nu}| \ll 1. \tag{4.110}$$

From Eq. (4.107) it can be shown that this implies

$$\Gamma^\mu_{00} = -\frac{1}{2}\eta^{\mu\beta}\frac{\partial h_{00}}{\partial x^\beta}. \tag{4.111}$$

In this way the time component ($\mu = 0$) of the equation (4.109) simply becomes $d^2x^0/ds^2 = 0$ since, by virtue of the assumption that h_{00} slowly changes with time (third condition), one can set $h_{00,0} \simeq 0$ on the right-hand side of Eq. (4.111), so that $\Gamma^0_{00} \simeq 0$. This implies $dx^0/ds = \text{const} = 1$. On the other hand the spatial components of Eq. (4.109) give

$$\frac{d^2 \vec{x}}{dt^2} \simeq -\frac{1}{2} c^2 \vec{\nabla} h_{00} . \tag{4.112}$$

Finally, identifying $h_{00} = 2\phi/c^2$, one recovers Newton's law. This final identification is a requirement imposed on the theory that is often referred to as the *correspondence principle*. In this way, we can see that general relativity can be indeed regarded as a generalisation of Newton's theory of gravity that is recovered under the three conditions defining the Newtonian limit.

However, we have still to show that also Newton's law of universal gravitation can be reproduced from general relativity and, to this extent, we will have still to turn to the correspondence principle.

4.3.5 Vectors and tensors in general relativity

The definition of a 4-vector in special relativity can be extended also to general relativity. We have seen that in general relativity one considers covariance of equations under general coordinate transformations $x^\mu \rightarrow x'^\mu(x^\nu)$.

A 4-vector (with contravariant index) is defined as a vector that transforms like

$$V'^\mu = V^\nu \frac{\partial x'^\mu}{\partial x^\nu} . \tag{4.113}$$

For a Lorentz transformation one recovers the definition of a 4-vector given in special relativity. The definition of tensors in general relativity proceeds analogously to what we discussed within special relativity.

4.3.6 The geometric analogy

We have seen that simple *global* coordinate transformations have the effect to change the metric tensor bringing it from the Minkowski metric tensor $\eta_{\mu\nu}$ to a metric tensor $g_{\mu\nu}$. This should be regarded as a sort of Minkowski metric tensor *in disguise*, i.e., completely equivalent to $\eta_{\mu\nu}$ from a physical point of view. We have seen, however, that the gravitational field is also described by a metric tensor $g_{\mu\nu}$. This is now the result of a *local* change of coordinates. Is there a way to distinguish whether a metric tensor $g_{\mu\nu}$ is either associated to a gravitational field or just to a change of coordinate system?

Given a metric tensor $g_{\mu\nu}$, one can build a rank-4 tensor $R^\mu_{\nu\alpha\beta}(g_{\mu\nu})$, the *Riemann–Christoffel curvature tensor* (or shortly the *Riemann tensor*), such that:[24]

---

[24]The expression $R^\mu_{\nu\alpha\beta}(g_{\mu\nu})$ is quite cumbersome; we again refer to general relativity text books such as [4, 9].

- $g_{\mu\nu}$ is a metric tensor in the *absence* of a gravitational field $\Leftrightarrow$ $R^{\mu}_{\nu\alpha\beta} = 0$,

- $g_{\mu\nu}$ is a metric tensor in the *presence* of a gravitational field $\Leftrightarrow$ $R^{\mu}_{\nu\alpha\beta} \neq 0$.

The Riemann tensor was originally introduced by Riemann to extend the concept of Gaussian curvature of the usual 2-dimensional surfaces embedded in the 3-dimensional space to a generic N-dimensional space. It can be calculated for 3-dimensional hyper-surfaces embedded in a 4-dimensional space and one discovers that it vanishes for a plane while it is non-zero for curved space. Therefore, we can also translate the previous statements in a geometrical language and write

- absence of gravitational field $\Leftrightarrow$ flat space,

- presence of gravitational field $\Leftrightarrow$ curved space.

This is why it is also possible to re-interpret the equation of motion of a free-falling massive point-like particle in a gravitational field as a *geodesic equation*: the particle moves in the curved space in a different way simply because it moves along the shortest path between two points, where the length of the path is measured by the interval s. Such paths are called *geodesics* and if the space is curved they do not correspond to straight lines.

Notice that a *geodesic equation* can also be written for massless particles like photons (though it has to be written in a different way since for photons $ds = 0$) and this implies that *light is also deviated by a gravitational field*.

4.3.7 Einstein's equations

We have now to find the equations for the gravitational field itself. In Newton's theory, the (gravitational) potential in a generic point $\vec{x}$ generated by a point-like particle of mass M located in the origin is given by

$$\phi(\vec{x}) = -\frac{G\,M}{|\vec{x}|}\,. \tag{4.114}$$

The potential generated in $\vec{x}$ by a generic distribution of massive point-like particles of masses M_i located in $\vec{x}_i$ will be the sum over all contributions,

$$\phi(\vec{x}) = -G \sum_i \frac{M_i}{|\vec{x} - \vec{x}_i|}\,. \tag{4.115}$$

If one has a continuous distribution of matter (a fluid) with mass density ρ, then the potential is determined by Poisson's equation

$$\nabla^2 \phi = 4\,\pi\,G\,\rho\,. \tag{4.116}$$

Let us concentrate on the right-hand side of this equation. How is it generalised in general relativity? We have again to start from special relativity. The mass density is not a scalar in special relativity but, as we have seen, it appears as the time-time component in the non-relativistic limit of the energy-momentum tensor for an ideal fluid. From Eq. (4.74) we can see explicitly that $\rho = T^{00}/c^2$. At the same time we have seen that in the Newtonian limit, obtained for non-relativistic motion and for a weak field, one has $\phi = c^2 h_{00}/2$. In this way we can recast Poisson's equation as

$$\nabla^2 h^{00} = \frac{8\pi G}{c^4} T^{00}, \qquad (4.117)$$

that, in the non-relativistic limit, gives the Newtonian gravitational field. Therefore, in general, we expect the gravitational field equations to be given by

$$\boxed{G^{\mu\nu} = \frac{8\pi G}{c^4} T^{\mu\nu}}. \qquad (4.118)$$

These are *Einstein's equations* for the gravitational field and the tensor $G^{\mu\nu}$ is the *Einstein tensor*.

Clearly the Einstein tensor has to be a function of the *metric tensor* $g_{\mu\nu}$ that, as we said many times, has to be identified with the gravitational field itself. Indeed the Einstein tensor is related to the curvature tensor (that, in turn, can be expressed in terms of $g_{\mu\nu}$ though we did not give its cumbersome expression) through the relation

$$G_{\mu\nu} = R_{\mu\nu} - \frac{1}{2} g_{\mu\nu} \mathcal{R}, \qquad (4.119)$$

where $R_{\mu\nu} \equiv R^{\lambda}{}_{\mu\lambda\nu}$ is the *Ricci tensor* and $\mathcal{R} \equiv R^{\mu}{}_{\mu}$ is the *curvature scalar*.

Applying the equivalence principle, in a similar way to what was done for the derivation of the geodesic equation, it can be shown that the *energy-momentum conservation equation* (4.75) in the presence of a gravitational field becomes

$$\boxed{T^{\mu}{}_{\nu;\mu} = 0}, \qquad (4.120)$$

where the symbol ";" indicates the operation of *covariant derivative*, defined as

$$T^{\mu}{}_{\nu;\mu} \equiv T^{\mu}{}_{\nu,\mu} + \Gamma^{\mu}{}_{\lambda\mu} T^{\lambda}{}_{\nu} - \Gamma^{\lambda}{}_{\nu\mu} T^{\mu}{}_{\lambda}. \qquad (4.121)$$

Before concluding this section, it is useful to summarise and highlight the results we have obtained. The gravitational field determines the geometry of space and this determines the motion of the fluid of particles with the geodesic equation. However in turn the distribution of particles determines, through Einstein's equations, the dynamics of the same gravitational field resulting into a coupled dynamics. This aspect of general relativity is nicely summarised by the words of John Wheeler: *Matter tells space how to curve [through Einstein's Equations] and space tells matter how to move [through the geodesic equation] [7].*

4.4 SYNCHRONOUS FRAMES

As we have seen, the metric tensor $g_{\mu\nu}$ plays in general relativity the same role played by the gravitational potential in Newton's theory. However, now one has to describe simultaneously the dynamics of all its ten independent components, a problem that in general seems quite challenging. It should be realised, however, that most of the complications arise because we are writing the metric tensor in a very general coordinate system. There is a first great simplification that one can introduce in different contexts including cosmology. Let us write the metric extensively, separating purely time, space and mixed components,

$$ds^2 = g_{\mu\nu}\, dx^\mu dx^\nu = g_{00}\, (dx_0)^2 + g_{0i}\, dx^0\, dx^i + g_{ij}\, dx^i\, dx^j\,. \tag{4.122}$$

In special relativity the first operation we had to do, in order to write equations in the simplest possible form, was to synchronise the watches. In general relativity a great source of complications in the equations comes from the fact that we are allowing watches to be asynchronous and we are even allowing their speed to vary with time. In our expanding universe, like in many other problems, it is convenient to work in systems where (coordinate) watches are synchronised.

Let us go back to the case of special relativity (no gravitational field) in order to understand what the properties of a synchronous metric look like. Consider the watch located in the origin O measuring the time t_O^{em} when a light signal is emitted and sent toward a point $\vec{x}_P$. Here another watch measures the signal arrival time t_P^{arrival}. The light signal is then reflected back to O where it arrives at t_O^{back}. The two watches are synchronised if

$$\Delta t_{O\to P} \equiv t_P^{\mathrm{arrival}} - t_O^{\mathrm{em}} = t_O^{\mathrm{back}} - t_P^{\mathrm{arrival}} \equiv \Delta t_{P\to O}\,. \tag{4.123}$$

If this condition is not verified, then one has to make a time correction (*synchronization*) on the watch located in P at $\vec{x}_P$ such that

$$t_P \to t_P' = t_P - \Delta t(\vec{x}_P)\,, \tag{4.124}$$

where the correction is given by

$$\Delta t(\vec{x}_P) = \frac{\Delta t_{O\to P} - \Delta t_{P\to O}}{2}\,. \tag{4.125}$$

Notice that the time correction depends in general on the point P.

After this operation one would obtain the metric in a synchronised system, $ds^2 = \eta_{\alpha\beta} dx^\alpha\, dx^\beta$. How would the metric tensor appear in such an asynchronous reference frame? Suppose that at a given point $\vec{x}$ the new time t' is related to t by the transformation

$$t \to t' = t + \Delta t(\vec{x})\,. \tag{4.126}$$

In this case one would get for the metric in the asynchronous frame

$$ds^2 = c^2\, dt'^2 - 2\, \frac{\partial \Delta t}{\partial x^i}\, dx^i\, dt' + \ldots \tag{4.127}$$

Notice that now terms $g_{0i} \neq 0$ appear in the metric tensor, so that time and spatial mixed terms appear in the metric.

One can also perform an even more contrived transformation, such that the hands of a watch, in the same point, change speed with time. This new time would not correspond to the proper time. This implies that if you make repeated measurements of some physical process (e.g., the lifetime at rest of some elementary particle or nucleus) they would give different results. Mathematically this would correspond to a transformation $t \to t'$ of the kind $t = f(t')$. This transformation would clearly generate a term $c^2 \dot{f}^2 dt'^2$ and therefore in the new coordinate asynchronous system one would have $g_{00} \neq 1$.

This means that, in the absence of gravitational fields, in synchronised frames, the metric tensor has $g_{00} = 1$ and $g_{0i} = 0$. The same result applies, more generally, in the presence of a gravitational field. Therefore, we can certainly greatly simplify our problem choosing to work in a *synchronous frame*, where the metric can be written, more simply, as

$$ds^2 = c^2 dt^2 - \gamma_{ij} dx^i dx^j . \tag{4.128}$$

The spatial tensor $\gamma_{ij} = -g_{ij}$ determines the *spatial metric* $d\ell^2$, the only part of the metric left to be determined in synchronous frames. We have now to understand what kind of conditions γ_{ij} has to satisfy in our homogeneous and isotropic universe.

Notice that the spatial metric describes the geometry of our space that has to be regarded as a 3-dim *hyper-surface* embedded in a 4-dim space-time. Notice also that for a point at rest one has $dx^i = 0$, so that $c^2 dt^2 = ds^2$ or $dt = ds/c \equiv d\tau$, the proper time we introduced in special relativity. Indeed this is the time measured by a free-falling watch and because of the equivalence principle, special relativity is recovered.[25]

EXERCISES

Exercise 4.1 *Prove that the scalar product of two 3-vectors is indeed a (non-relativistic) scalar.*

Exercise 4.2 *Given a vector $\vec{a}$ and a tensor T, prove that $b_i = \sum_k T_{ik} a_k$ are the three components of a 3-vector $\vec{b}$.*

Exercise 4.3 *Prove that given two vectors $\vec{a}$ and $\vec{b}$, their direct product $T_{ij} = a_i b_j$ is a tensor.*

Exercise 4.4 *Prove that the cross product of two vectors $\vec{a}$ and $\vec{b}$ can be regarded both as an anti-symmetric and as a pseudo- (or axial-)vector.*

---

[25]On Earth we are not in free fall. This implies that our everyday reference frame is asynchronous. In particular for two observers located at different heights, with identical watches, time flows at different speeds.

Exercise 4.5 *Consider a Lorentz boost in the x direction. Show that this can be recast as a usual rotation in the plane $ix^0 - x$ (where $ix^0 = ict$ is the imaginary time) with a complex angle ψ'. How is ψ' related to ψ defined in Eq. (4.30)?*

Exercise 4.6 *Re-derive from Lorentz transformations the dilation of time effect for the life time of a particle in flight t in terms of the life time at rest.*

Exercise 4.7 *Show that the partial derivative $\partial/\partial x^\mu \equiv \partial_\mu$ transforms as a covariant 4-vector.*

Exercise 4.8 *Show that the matrix Λ, associated to a generic Lorentz transformation, has to satisfy the condition (4.46).*

Exercise 4.9 *Consider a coordinate transformation from an inertial system $\mathcal{S}$ of Cartesian coordinates (t, x, y, z) to a non-inertial system of coordinates (t, x', y', z') in uniform rotation with angular frequency ω and whose rotation axis $\hat{z}'$ coincides with $\hat{z}$ (therefore $z' = z$). Write down the coordinate transformations. Express the metric ds^2 in this new system and write down the metric tensor. Notice that the time coordinate is assumed to be the same in the two reference frames $(t' = t)$.*

Exercise 4.10 *Derive the expression (4.107) for the Christoffel symbols in terms of the metric tensor.*

Exercise 4.11 *Derive the energy-momentum tensor conservation equation (4.120) and the expression for the covariant derivative Eq. (4.121).*

BIBLIOGRAPHY

[1] For an extensive discussion on classical mechanics, see V.I. Arnold, *Mathematical methods of classical mechanics* (1974) Springer-Verlag, New York Inc.

[2] G. Galiliei, *Dialogue concerning the two chief world systems*, Florence, 1632.

[3] A. Einstein, *On the electrodynamics of moving bodies*, Annalen Phys. **17** (1905) 891 [Annalen Phys. **14** (2005) 194].

[4] L. D. Landau, E.M. Lifshitz, *The classical theory of fields*, Course of theoretical physics, Vol. 2, Pergamon press, Oxford, 1971.

[5] A. Einstein, *The foundation of the general theory of relativity*, Annalen Phys. **49** (1916) 769 [Annalen Phys. **14** (2005) 517].

[6] S. Weinberg, *Gravitation and cosmology: Principles and applications of the general theory of relativity*, John Wiley and Sons, January 1972.

[7] J. A. Wheeler, *Gravitation and space-time: The four-dimensional event space of relativity theory (a journey into gravity and space-time)*, Freeman, New York, 1990.

CHAPTER 5

Geometry of the universe

In this chapter we start applying general relativity to cosmology, deriving the
metric of a homogeneous and isotropic universe leading to Friedmann cosmological models. In general one would have to solve a complicated set of 10 differential equations (since tensors in Einstein's equations are symmetric) to determine
the dynamics of the metric. However, the cosmological principle greatly simplifies
the problem, since it reduces the metric to a very simple form, the Friedmann–
Robertson–Walker metric, depending just on one dynamical parameter, the scale
factor, and one discrete constant parameter, the curvature parameter fixing the
geometry of the universe. We present a simple geometric derivation, starting from
an insightful two-dimensional analogy.

5.1 COSMOLOGY FOR ANTS: THE TWO-DIMENSIONAL ANALOGY

We need to understand what conditions have to be imposed on the metric in order
to satisfy the cosmological principle (homogeneity and isotropy). This is equivalent
to imposing that there is no privileged point in space and no privileged direction
around any point.

A useful analogy is to consider first a usual two-dimensional surface embedded in
an ordinary three-dimensional space. This means that everything has to be assumed
to be confined to move on this surface, including light.

What are the two-dimensional surfaces respecting the cosmological principle?
The surface of a sphere first comes to mind, since it clearly has no privileged point
or direction. We have then to understand how to derive its metric tensor. The first
step is to imagine that this is embedded in a flat three-dimensional space whose
metric is nothing else than the usual Euclidean metric,[1]

$$dl^2 = dx_1^2 + dx_2^2 + dx_3^2 . \tag{5.1}$$

Any point on the surface of a sphere of radius R obeys the parametric condition

$$x_1^2 + x_2^2 + x_3^2 = R^2 . \tag{5.2}$$

---

[1]Notice that we can now skip the distinction between indexes up and down since we are dealing
only with the spatial part of the metric.

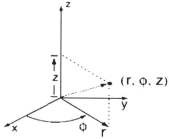

FIGURE 5.1: Cylindrical coordinates (from *Hyperphysics*, C.R. Nave).

This allows to express one of the three coordinates in terms of the other two. All points on the surface of a sphere have to be described by just two independent coordinates since it is a two-dimensional surface. For example, we can express x_3 in terms of x_1 and x_2. We can then first write

$$x_3^2 = R^2 - x_1^2 - x_2^2 \,, \tag{5.3}$$

and then, differentiating Eq. (5.2), we obtain

$$2\,x_1\,dx_1 + 2\,x_2\,dx_2 + 2\,x_3\,dx_3 = 0 \,. \tag{5.4}$$

In this way we can express dx_3^2 in Eq. (5.1) as

$$dx_3^2 = \frac{(x_1\,dx_1 + x_2\,dx_2)^2}{R^2 - x_1^2 - x_2^2} \,. \tag{5.5}$$

At this stage we would have already obtained the metric tensor in Cartesian coordinates but clearly this is not the simplest coordinate frame to write the metric tensor, since the rotational symmetry is not manifest.

Let us see how the metric reads in cylindrical polar coordinates (see Fig. 5.1). We can express x_1 and x_2 as

$$x_1 = \bar{r}\cos\phi \,, \tag{5.6}$$
$$x_2 = \bar{r}\sin\phi \,. \tag{5.7}$$

Differentiating, we obtain

$$dx_1 = \cos\phi\,d\bar{r} - \bar{r}\sin\phi\,d\phi \,, \tag{5.8}$$
$$dx_2 = \sin\phi\,d\bar{r} + \bar{r}\cos\phi\,d\phi \,. \tag{5.9}$$

From these expressions one can directly obtain dx_3^2 in Eq. (5.5) in cylindrical polar coordinates. There is a convenient shortcut that is also insightful. Since the surface of a sphere has no privileged point or direction, the spatial metric $d\ell^2$ has to be the same at any point.

Therefore, one can just calculate the metric in the most convenient point that, for a fixed $\bar{r}$, is that one for $\phi = 0$. In this case one simply has

$$x_1 = \bar{r}, \quad x_2 = \bar{r}\sin\phi = 0, \quad dx_1 = d\bar{r}, \quad dx_2 = \bar{r}\,d\phi. \tag{5.10}$$

and from Eq. (5.5) one obtains straightforwardly first

$$dx_3^2 = \frac{\bar{r}^2\,d\bar{r}^2}{R^2 - \bar{r}^2}, \tag{5.11}$$

and then, replacing $\bar{r} \to r \equiv \bar{r}/R$,

$$dx_3^2 = R^2\,\frac{r^2\,dr^2}{1 - r^2}, \tag{5.12}$$

obtaining finally

$$d\ell^2 = R^2\left(\frac{dr^2}{1 - r^2} + r^2\,d\phi^2\right). \tag{5.13}$$

The surface of the sphere has positive constant Gaussian curvature $K = 1/R^2$, as imposed by homogeneity. Moreover the angles of a triangle drawn on a sphere, obtained connecting three points with the geodesics, obey the relation

$$\alpha + \beta + \gamma = \pi + \frac{A}{R^2}, \tag{5.14}$$

where A is the enclosed area.

Analogously one can think of another surface respecting homogeneity and isotropy, characterised again by constant curvature but *negative*, $K = -1/R^2$. This surface is the *hyperbolic plane* and it provides a second case of homogeneous and isotropic two-dimensional surface, despite the fact that it cannot be embedded in a three-dimensional Euclidean space and, for this reason, it cannot be visualised or drawn. However, the region around a saddle point has approximately constant negative curvature and helps intuition.

All properties of the hyperbolic plane can be easily deduced from the surface of a sphere upon the replacement $R \to iR$. For example, the sum of the angles of a triangle drawn on the hyperbolic plane will be given by

$$\alpha + \beta + \gamma = \pi - \frac{A}{R^2} \tag{5.15}$$

and all points on the hyperbolic plane satisfy the relation

$$x_1^2 + x_2^2 + x_3^2 = -R^2, \tag{5.16}$$

where now also x_3 has to be necessarily purely imaginary. With the replacement $x_3 \to i\,x_3'$, one has

$$x_1^2 + x_2^2 - x_3'^2 = -R^2, \tag{5.17}$$

where now x'_3 is real.

Starting from this parametric equation defining the hyperbolic plane and repeating the derivation done for the surface of a sphere, in particular expressing x'_3 in terms of x_1 and x_2, one eventually obtains for the spatial metric

$$d\ell^2 = R^2 \left(\frac{dr^2}{1 + r^2} + r^2\, d\phi^2 \right). \tag{5.18}$$

This metric defines the *hyperbolic* (or *Lobachevsky*) geometry.

A flat space ($K = 0$), that in our two-dimensional analogy reduces to an infinite plane, is another clear case of a homogeneous and isotropic surface, with everywhere vanishing curvature. It can be clearly regarded as a limit of the previous two cases for $R \to \infty$. In this case the metric in polar coordinates is found quite easily[2]

$$d\ell^2 = R^2 \left(dr^2 + r^2\, d\phi^2 \right). \tag{5.19}$$

Introducing the *curvature parameter k*, we can write in a compact way the metric for all three cases as

$$d\ell^2 = R^2 \left(\frac{dr^2}{1 - k\, r^2} + r^2\, d\phi^2 \right), \tag{5.20}$$

where $k = -1, 0, +1$ correspond respectively to hyperbolic plane, plane and sphere.

Notice that the hyperbolic plane is an unbounded surface with infinite area. Therefore, it describes a so-called *open space* contrarily to the sphere that describes a *closed space*. The sphere has the peculiar property to be unbounded but with a finite volume. Finally, the plane is the two-dimensional realisation of a *flat space*.

5.2   THE FRIEDMANN–ROBERTSON–WALKER METRIC

We can now extend the results found for two-dimensional spaces to a three-dimensional space. Specifically, we want to find again the geometries that satisfy the cosmological principle and whose properties are, therefore, independent of the position of the observer and of the direction of observation around the point.

Again this implies that we have to look for those three-dimensional spaces where *curvature is constant*. As mentioned in the previous chapter, in a generic three-dimensional space, curvature is described by the Riemann–Christoffel tensor $R^\lambda_{\rho\sigma\nu}$. In particular a space with constant curvature is a space where $\mathcal{R}$, the curvature scalar, is constant.

This generalises Gaussian curvature to a space with a generic number of dimensions. Again the first example coming to mind, using the two-dimensional analogy, is the three-dimensional sphere. We have to repeat the same steps, with a proper

---

[2]One could worry whether the metric is divergent since $R \to \infty$. However, at the same time $r = \bar{r}/R \to 0$ in the denominator and one has simply $d\ell^2 = d\bar{r}^2 + \bar{r}^2\, d\phi^2$: the divergence is only apparent.

generalisation, discussed in the two-dimensional case. Let us again imagine embedding the 3-dim sphere in a 4-dim space (the 3-dim sphere is then an *hypersphere* in this space) with Euclidean metric

$$d\ell^2 = dx_1^2 + dx_2^2 + dx_3^2 + dx_4^2 \,. \tag{5.21}$$

The hypersphere will be then described by the parametric equation

$$x_1^2 + x_2^2 + x_3^2 + x_4^2 = R^2 \,. \tag{5.22}$$

At this point, analogously to what we did in two dimensions, we can express the x_4 coordinate in terms of the other three, writing

$$x_4^2 = R^2 - (x_1^2 + x_2^2 + x_3^2) \,, \tag{5.23}$$

and differentiating one obtains

$$dx_4^2 = \frac{(x_1 \, dx_1 + x_2 \, dx_2 + x_3 \, dx_3)^2}{R^2 - x_1^2 - x_2^2 - x_3^2} \,. \tag{5.24}$$

In 3-dim the role that in 2-dim was played by *polar-cylindric* coordinates (see Fig. 4.1), is now played by the spherical coordinates in terms of which

$$
\begin{aligned}
x_1 &= \bar{r} \sin\theta \cos\phi \\
x_2 &= \bar{r} \sin\theta \sin\phi \\
x_3 &= \bar{r} \cos\theta \,.
\end{aligned}
\tag{5.25}
$$

We already wrote this transformation (see Eq. (4.89)), now we are simply denoting the radial dimensionful coordinate with $\bar{r}$ instead of r. Starting from a flat space we obtained Eq. (4.91) for the metric. This time the same result Eq. (4.91) will hold only for part of the metric, so that we can write

$$dx_1^2 + dx_2^2 + dx_3^2 = d\bar{r}^2 + \bar{r}^2 \, [d\theta^2 + \sin^2\theta \, d\phi^2] \,. \tag{5.26}$$

We also have the additional term dx_4^2 as a consequence of studying a curved 3-dim space. This term can be calculated exactly as we did for the 2-dim sphere in polar-cylindric coordinates. We used the trick to calculate dx_1 and dx_2 around points with $\phi = 0$. In this case we calculate the three dx_i's around points with $\theta = \phi = 0$. The result for dx_4^2 is exactly the same as we were obtaining in the two-dimensional case for dx_3^2, namely

$$dx_4^2 = \frac{\bar{r}^2 \, d\bar{r}^2}{R^2 - \bar{r}^2} \,. \tag{5.27}$$

Therefore, combining everything together, we finally obtain

$$d\ell^2 = R^2 \left(\frac{dr^2}{1 - r^2} + r^2 \, d\Omega^2 \right) \,, \tag{5.28}$$

where we have again replaced $\bar{r} \to r \equiv \bar{r}/R$ and defined $d\Omega^2 \equiv d\theta^2 + \sin^2 \theta \, d\phi^2$.

We can also again extend the result to a space with *constant negative curvature* and to a *flat space* obtained for $R \to \infty$ and we can again express all three cases in a compact way introducing the curvature parameter $k = -1, 0, +1$, writing

$$d\ell^2 = R^2 \left(\frac{dr^2}{1 - k \, r^2} + r^2 \, d\Omega^2 \right). \tag{5.29}$$

This is the *spatial metric* of a homogeneous and isotropic universe.[3]

The last step is to go back to the 4-dim *space-time* and write finally

$$ds^2 = c^2 \, dt^2 - R^2(t) \left(\frac{dr^2}{1 - k \, r^2} + r^2 \, d\Omega^2 \right). \tag{5.30}$$

This is the famous *Friedmann–Robertson–Walker (FRW) metric* [1, 2]. Notice that in general the radius of curvature will be a function of time, $R = R(t)$. If we indicate with R_0 the value of R at present, then we can recast the FRW metric as

$$\boxed{ds^2 = c^2 \, dt^2 - a^2(t) \, R_0^2 \left(\frac{dr^2}{1 - k \, r^2} + r^2 \, d\Omega^2 \right),} \tag{5.31}$$

where $a(t) \equiv R(t)/R_0$ is the *scale factor*, a quantity that we have already introduced in the case of the Milne–McCrea model but that now has quite a different physical meaning.[4]

In the case of *flat space* one could legitimately object that, since $R_0 \to \infty$, Eq. (5.31) is meaningless. This is true and indeed in this case R_0 should be regarded not as the radius of curvature but as a conveniently chosen reference length, for example a convenient unit of measurement of lengths such as 1 Mpc, the typical inter-galactic distance. In any case this ambiguity is harmless, since R_0 will always cancel out and one is left with just the scale factor a.

Remember that we are writing our metric in a synchronous system, such that reference (free-falling) watches measure the proper time and are synchronous with each other. In addition the system is not only synchronous but also *comoving*: coordinate points move jointly with free massive particles at rest (matter). We will discuss this important aspect in greater detail in the next chapter.

The variation of the scale factor $a(t)$ with time, results in a rescaling of all physical distances between points with fixed comoving coordinates, i.e., at rest in the comoving system. If we define the *comoving spatial metric element* as

$$d\ell_{(0)} = R_0 \sqrt{\left(\frac{dr^2}{1 - k \, r^2} + r^2 \, d\Omega^2 \right)}, \tag{5.32}$$

---

[3]One could wonder whether in 3-dim one could have additional cases beyond these three geometries with constant curvature, but it can be shown that they cover all possibilities.

[4]Notice that the scale factor is normalised in a way that at the present time $a_0 = 1$.

one has

$$dl = a(t)\, dl_{(0)}.$$ (5.33)

This relation shows that if the comoving spatial element is constant, then the physical spatial element scales increases $\propto a(t)$. The relation (5.33) can also be integrated on a finite size along some suitably defined curvilinear coordinate (we will show some examples). In this way a physical scale λ is related to the corresponding comoving scale $\lambda_{(0)}$ by the simple relation

$$\lambda(t) = a(t)\, \lambda_{(0)}(t).$$ (5.34)

For points at rest in the comoving system one has $\lambda_{(0)} = $ const and the physical distance $\lambda \propto a(t)$. For example, the comoving distance of two far unbound clusters of galaxies at rest in the comoving system is by definition constant and therefore in this case their physical distance increases linearly with $a(t)$. We will be back on this point in the next chapter.

On the other hand, if one has a *bound* physical system whose objects are not free falling but interacting with each other strongly enough that $\lambda = $ const, then consequently $\lambda_{(0)} \propto a^{-1}$. Examples of bound systems are the human body, planets, stars, the solar system, galaxies. Therefore, for all these systems the comoving size decreases with the expansion. Clearly the existence of such bound systems implies that locally there can be a departure from a perfectly homogeneous universe. This is not at all in contradiction with the cosmological principle that, as we said, applies only on scales larger than 100 Mpc. Therefore, the FRW metric has to be regarded as a background metric rigorously valid on such large scales where matter is at rest in the comoving system and is not gravitationally bound. On smaller scales one has inhomogeneities and formation of bound systems such as galaxies and clusters of galaxies. On such scales matter acquires peculiar velocities, i.e., velocities with respect to the comoving system.

5.3 TOPOLOGY OF THE UNIVERSE

Imposing that the cosmological principle applies to all points of the universe, we obtained three very well-defined spaces: the spherical space, the flat space and the hyperbolic plane. In the first case the universe has a finite size, while in the other two it has an infinite size.

This conclusion relies on the possibility that we can apply the cosmological principle, that as discussed is supported by the cosmological observations, to the whole universe. However, if the universe that we observe is only just a portion. of size λ_{obs}, of the whole universe, this assumption can well break down at scales larger than λ_{obs}.

In the case of a closed spherical universe it is quite simple to conclude that the observable portion of the universe (it is simply called the *observable universe*) cannot be the entire 3-sphere with radius R.[5] Suppose indeed that on the contrary the size of our observable universe $\lambda_{\mathrm{obs}} > \pi R$. This would correspond to a situation where the cosmological observations are able to encompass the whole universe. Light, like massive particles, moves on geodesics of the space-time and therefore in a curved space it would move along curved lines.

In the case of a closed spherical universe with $\lambda_{\mathrm{obs}} > \pi R$ we would then be able to have multiple images of the same object (e.g., the same galaxy) from different directions! In principle we would even be able to observe Earth itself but in the past, at a time $\Delta t = 2\pi R/c$ ago. However, astronomical observations certainly exclude this fascinating situation. but it does not mean that they rule out a closed spherical universe.

This is because it can well be that simply our observable universe is just a tiny fraction of the entire 3-sphere, so that $\lambda_{\mathrm{obs}} \ll R$. Notice that in the limit $\lambda_{\mathrm{obs}}/R \to 0$ the case of a flat universe is recovered anyway. This is quite clear since for a 2-dim curved surface, a sufficiently small portion around any given point can always be approximated by its tangent plane. For example, observations certainly do not show any evidence of curvature on scales as small as the solar system. We would, therefore, expect to be able to see any curvature effect only on much larger scales.

Actually we will see that λ_{obs} is indeed upper bounded (it is not infinite), meaning that we do not have any information on points at distances larger than λ_{obs}. We will also see that, even on scales as large as λ_{obs}, all experimental observations do not show any evidence of curvature, i.e., the geometry of our observable universe is indistinguishable from that of a flat space. This seems that we are really in a situation similar to that of ancient populations that had access only to a limited portion of Earth surface and were unable to detect the non-vanishing Earth curvature: history seems to repeat from this point of view.

Therefore, it seems that our observable universe is only a small portion of the whole universe and that it makes sense to apply the cosmological principle only to this portion. Current observations do not give us any information at the moment about the *global shape of the universe* or, more technically, about the *topology of the universe*. Later on we will discuss how the model of inflation gives a precise explanation of the reason why our observable universe looks flat and how it predicts our observable universe to be just a tiny portion of the whole universe. This implies that, at least in a conventional 3-dim spatial universe, there is no hope to test global properties of the universe.

There are, however, speculative cosmological models, within theories with extra spatial dimensions, where in principle we could get information on regions beyond our observable universe. They would be far in our usual 3-dim space but close in a hypothetic fourth extra-dimension and we could hope to get information on these far regions through so-called *short-cuts* in space-time. This kind of 4-dim space

---

[5]As we will discuss in Chapter 7, this possibility is realised by the static Einstein's universe, the first proposed cosmological model within general relativity.

topology has also been invoked to explain the dark matter of the universe if one allows gravitation to propagate through the extra-dimension. In this case the dark matter would be nothing else than very ordinary matter along the usual dimensions but close in the fourth dimension, a sort of parallel world. All these ideas can be formulated in a rigorous way but within theories and models beyond our scope.

It is therefore time to go back to our 3-dim observable universe and rigorously work out all physical properties implied by general relativity and in particular by the FRW metric derived in this chapter. The two-dimensional analogy will provide a useful pictorial tool for an intuitive understanding of geometrical properties of three-dimensional spaces.

EXERCISES

Exercise 5.1 *Suppose you are an ant living on the surface of a sphere with radius R. An object with size $d\lambda \ll R$ is at distance ℓ from you (all distances are measured on the surface of the sphere). What angular width $d\psi$ will you measure for the object? Explain the behavior of $d\psi$ as $\ell \to \pi R$.*

Exercise 5.2 *Show that if you draw a circle of radius ℓ on a spherical surface, the circle's circumference is going to be $C = 2\pi R \sin(\ell/R)$. Imagine one can measure distances with an error ± 1m, how long a circle would you have to draw on the Earth's surface ($R = 6371$ km) to convince yourself that the Earth is spherical rather than flat?*

Exercise 5.3 *Consider the 3-dim spatial metric*

$$d\ell^2 = \frac{d\bar{r}^2}{1 - k\,\bar{r}^2/R^2} + \bar{r}^2\,d\Omega^2\,. \tag{5.35}$$

Knowing that with the coordinate transformation $\bar{r} = S_k(x)$ the metric gets expressed as

$$d\ell^2 = dx^2 + S_k^2(x)\,d\Omega^2\,, \tag{5.36}$$

find $S_k(x)$ for $k = +1, 0, -1$. What is the physical meaning of the new coordinate x?

Exercise 5.4 *The physical size of the Galaxy is approximately 25 kpc. What is its comoving size at present? What was its comoving size when $a(t) = 0.2$?*

BIBLIOGRAPHY

[1] A. Friedman, *On the curvature of space*, Z. Phys. **10** (1922) 377 [Gen. Rel. Grav. **31** (1999) 1991].

[2] H. P. Robertson, *Kinematics and world-structure*, Astrophysical Journal, **82** (1935) 284; H. P. Robertson, *Kinematics and world-structure II*, Astrophysical Journal, **83** (1936) 187; H. P. Robertson, *Kinematics and world-structure*

III, Astrophysical Journal, **83** (1936) 257; A. G. Walker, *On Milne's theory of world-structure*, Proceedings of the London Mathematical Society, **s2-42** (1937) 90.

CHAPTER 6

Dynamics of the universe

In this chapter we will work out various interesting physical consequences of the FRW metric and the importance of the comoving system that should be regarded as a unique privileged system to describe the behaviour of matter and radiation. In particular, we will discuss how cosmological redshifts are related to the cosmological expansion. We will then define the proper distance of an astronomical object in the expanding universe and how this relates to the luminosity distance. In this way we will finally be able to derive Hubble's law on pure theoretical grounds. Finally we will see how Friedmann's equation and the fluid equation, that we could only partly justify within a Newtonian approach, can be obtained in a rigorous way from Einstein's equations starting from the FRW metric.

6.1 THE COMOVING SYSTEM

Let us discuss some remarkable physical properties of the comoving coordinate system,[1] that should make clear how this is not just a convenient coordinate system where equations acquire particularly simple expressions, but should be regarded as the natural reference frame to describe physical processes and such that the symmetries of the universe, homogeneity and isotropy, are explicitly manifest. In particular velocities and momenta of massive objects (from individual elementary particles to galaxies) can be unambiguously calculated, from a cosmological point of view, only in the comoving system.

6.1.1 Absolute scale of velocities

The cosmological principle holds only in the comoving system. In a boosted reference system, in motion with respect to the comoving system, there would be an apparent privileged direction originating by the boost itself. This effect has been experimentally tested, since the motion of the Earth in the comoving system induces a dipole anisotropy in the CMB radiation that has been observed [1]. This means that the CMB temperature of the forward sky hemisphere in the direction of

---

[1]We will shortly refer to it as just the *comoving system*.

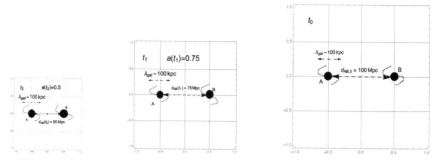

FIGURE 6.1: Proper distance between two galaxies A and B at rest in the comoving system at three different times $t_2 < t_1 < t_0$ such that $a(t_2) = 0.5$, $a(t_1) = 0.75$ and $a_0 = 1$. The grid indicates *comoving coordinates* in units of $100\,\mathrm{Mpc}$, clearly showing how the comoving distance remains constant: $d_{AB}^{(0)}(t_2) = d_{AB}^{(0)}(t_1) = d_{AB,0}^{(0)} = 100\,\mathrm{Mpc}$. On the other hand the physical size of the two galaxies, $\lambda_{\mathrm{gal}} \sim 100\,\mathrm{kpc}$ (it is not in scale!), remains constant while their comoving size shrinks with time.

Earth's velocity is different from the temperature of the backward sky hemisphere. Notice that this observation does not imply a violation of the principle of relativity and Lorentz invariance, since in this case one is detecting a boost of a part with respect to the whole system. Lorentz invariance is supposed to hold if ideally one could perform a boost of the whole universe itself.

6.1.2 Free-falling objects initially at rest in the comoving system remain at rest forever

If a free-falling massive object (i.e., subject only to the cosmic gravitational field, with negligible additional interactions) is at some initial time t_{in} at rest in the comoving system (it means that at t_{in} one has $dr = d\theta = d\phi = 0$), it will remain at rest forever. This means that comoving coordinate points track the motion of free-falling bodies (and vice versa). Using a 2-dim analogy, one should think of (bi-dimensional) objects at rest on the surface of an expanding balloon.

Notice that this result implies that the *physical distance* between two free-falling objects (e.g., two galaxies) A and B at rest in the comoving system scales proportionally to the scale factor[2]

$$d_{AB}(t) = a(t)\, d_{AB,0}. \tag{6.1}$$

Fig. 6.1 shows a schematic representation of how the proper distance between two far galaxies at rest in the comoving system changes between a past time t_2 when the scale factor was halved and the present time. We will see in a moment that this distance generalises the definition of *proper distance* discussed in special relativity. Notice that Eq. (6.1) extends the infinitesimal expression Eq. (5.33) to the finite case. We will see how to calculate explicitly proper distances.

---

[2]On the other hand their comoving distance $d_{AB}^{(0)}(t) = d_{AB}^{(0)}(t_0) = d_{AB}(t_0)$ is of course constant.

6.1.3 Redshift of momentum

More generally, if a free-falling massive object is not at rest in the comoving system but has momentum $\vec{p}_{in}$ at some initial time t_{in}, it will afterwards lose momentum (and therefore velocity) as

$$|\vec{p}|(t) = |\vec{p}|_{in} \frac{a_{in}}{a(t)} \propto a(t)^{-1}, \tag{6.2}$$

where $a_{in} \equiv a(t_{in})$. This result is often referred to as *redshift of momentum*.[3] It clearly implies that $\gamma \, |\vec{v}| \propto a^{-1}$ and, therefore, that massive objects lose velocity and asymptotically tend to be at rest in the comoving system (if, as we are assuming, all interactions with the other parts of the system are negligible, including local gravitational ones, neglected in our smoothed universe).

---

[3]Redshift of momentum follows from the geodesic equation (see Eq. (4.105)) given by

$$\frac{d^2 x^\mu}{ds^2} = -\Gamma^\mu_{\alpha\beta} \frac{dx^\alpha}{ds} \frac{dx^\beta}{ds}. \tag{6.3}$$

This can be recast, in terms of the 4-velocity $u^\mu \equiv dx^\mu/ds = \gamma \, (1, \vec{v}/c)$, as

$$\frac{d u^\mu}{ds} = -\Gamma^\mu_{\alpha\beta} u^\alpha u^\beta. \tag{6.4}$$

Let us now consider the spatial components of the equation of the geodesics (remember that these are indicated with Latin indexes $i = 1, 2, 3$)

$$\frac{d u^i}{ds} = -\Gamma^i_{\alpha\beta} u^\alpha u^\beta. \tag{6.5}$$

Out of the 16 terms on the r.h. side, only terms with $\alpha = 0$ and $\beta = i$ do not vanish and since $\Gamma^i_{0i} = \dot{a}/a$ one has

$$\frac{d u^i}{ds} = -\gamma \frac{\dot{a}}{a} u^i, \tag{6.6}$$

where we wrote $u^0 = \gamma$. Now one can always choose the spatial coordinates such that the velocity has only one non-zero component that in this case has to coincide with $|\vec{u}|$. Therefore, the last equation has to hold for $|\vec{u}|$ as well and we can write

$$\frac{d|\vec{u}|}{ds} = -\gamma \frac{\dot{a}}{a} |\vec{u}|. \tag{6.7}$$

Notice now that the infinitesimal spatial element is given by $d\ell = |\vec{v}| \, dt$ and therefore $ds^2 = c^2 \, dt^2 - |\vec{v}|^2 \, dt^2 = dt^2/\gamma^2$ and therefore $ds = dt/\gamma$. In this way we can finally write

$$\frac{1}{|\vec{u}|} \frac{d|\vec{u}|}{dt} = -\frac{\dot{a}}{a}. \tag{6.8}$$

Since the 4-momentum is given by $p^\mu = m \, c \, u^\mu$ and therefore $\vec{p} = m \, c \, \vec{u}$, we can then also write

$$\frac{1}{|\vec{p}|} \frac{d|\vec{p}|}{dt} = -\frac{\dot{a}}{a}, \tag{6.9}$$

and from this one finally obtains

$$|\vec{p}|(t) \propto a(t)^{-1}. \tag{6.10}$$

This explains why galaxy peculiar velocities, generated by the galaxy local inter-actions due to inhomogeneities on scales smaller than $\sim 100\,\mathrm{Mpc}$, have maximum values $\mathcal{O}(1000\,\mathrm{km\,s^{-1}})$ due to local gravitational inhomogeneities.

6.1.4 Cosmological redshifts

Redshift of momentum applies to photons as well.[4] However, since the universality of the speed of light (holding in special relativity) has to be valid also in the presence of gravitational fields (it is a straightforward consequence of the principle of equivalence), in the case of photons the redshift of momentum does not imply that photons slow down during the expansion. They will never end up at rest in the comoving system, as in any other system.

Considering that $|\vec{p}| = \hbar/\lambda$, where λ is the electromagnetic wavelength in the case of photons and the De Broglie wavelength in the case of massive particles, redshift of momentum implies

$$\lambda(t) = \lambda_0 \frac{R(t)}{R_0} = a(t)\,\lambda_0\,. \tag{6.11}$$

This means that wavelengths scale proportionally to $a(t)$, exactly like proper distances between galaxies at rest in the comoving system. Pictorially, this can be expressed saying that particle wavelengths are stretched by the universe expansion.

This result implies that the redshift of a luminous signal, defined by Eq. (2.19), emitted at a time t_{em} by a source at rest in the comoving system and detected at the present time t_0, is simply given by

$$z \equiv \frac{\lambda_0}{\lambda_{\mathrm{em}}} - 1 = \frac{R_0}{R_{\mathrm{em}}} - 1 = a_{\mathrm{em}}^{-1} - 1\,, \tag{6.12}$$

where we defined $\lambda_{\mathrm{em}} \equiv \lambda(t_{\mathrm{em}})$, $R_{\mathrm{em}} \equiv R(t_{\mathrm{em}})$ and $a_{\mathrm{em}} \equiv a(t_{\mathrm{em}})$. Notice in particular that $z \to 0$ for $t_{\mathrm{em}} \to t_0$. It is quite instructive to see how this important result is also valid classically, when light is described in terms of electromagnetic waves.

As we said, the principle of equivalence implies that the speed of light is a universal constant also in the presence of a gravitational field. Therefore, as in special relativity, during light propagation one has $ds = 0$. Consider now an electromagnetic wave crest emitted at t_{em} by a luminous source at rest in the comoving system. This is detected at the present time t_0 by an observer on the Earth (assumed to be approximately at rest in the comoving system) conveniently set as the origin of the system. In this way during the light signal propagation one has simply $d\theta = d\phi = 0$. Therefore, from the expression for the FRW metric Eq. (5.30), one can write

$$\frac{c\,dt}{R(t)} = -\frac{dr}{\sqrt{1 - k\,r^2}}\,, \tag{6.13}$$

where the minus sign takes into account that the luminous signal is propagating

---

[4]This can be shown, analogously to the case of massive particles discussed in the previous footnote, starting from the geodesics equation for massless particles.

from the source to us, so that $dr < 0$. If we now integrate between t_{em} and t_0, we obtain

$$\int_{t_{em}}^{t_0} \frac{c\,dt}{R(t)} = \int_0^{r_{em}} \frac{dr}{\sqrt{1 - k\,r^2}} \equiv f(r_{em}), \tag{6.14}$$

where notice that the second integral is a function only of r_{em}. Therefore, considering the emission of the subsequent wave-crest at $t_{em} + \Delta t$, where $\Delta t = \lambda/c$ is the oscillation period of the e.m. wave, we have

$$\int_{t_{em}+\lambda_{em}/c}^{t_0+\lambda_0/c} \frac{c\,dt}{R(t)} = \int_0^{r_{em}} \frac{dr}{\sqrt{1 - k\,r^2}} \equiv f(r_{em}), \tag{6.15}$$

where r_{em} is unchanged, since we are assuming that the source is at rest in the comoving system. The first integral can be decomposed in three terms, explicitly:

$$\int_{t_{em}}^{t_0} \frac{c\,dt}{R(t)} + \int_{t_0}^{t_0+\lambda_0/c} \frac{c\,dt}{R(t)} - \int_{t_{em}}^{t_{em}+\lambda_{em}/c} \frac{c\,dt}{R(t)} = \int_0^{r_{em}} \frac{dr}{\sqrt{1 - k\,r^2}} \equiv f(r_{em}). \tag{6.16}$$

Taking the difference, side to side, between this last equation and Eq. (6.14), we find

$$\int_{t_0}^{t_0+\lambda_0/c} \frac{c\,dt}{R(t)} - \int_{t_{em}}^{t_{em}+\lambda_{em}/c} \frac{c\,dt}{R(t)} = 0. \tag{6.17}$$

Since during the period of oscillation the variation of $R(t)$ is completely negligible, we obtain finally

$$\frac{\lambda_0}{R_0} = \frac{\lambda_{em}}{R_{em}} \quad \Rightarrow \quad \frac{\lambda_0}{\lambda_{em}} = \frac{R_0}{R_{em}}. \tag{6.18}$$

This derivation clearly shows how the cosmological redshift should not be regarded as a Doppler effect but as a genuine geometrical effect due to the space expansion that stretches the wavelength of the electromagnetic waves while they propagate.

6.2 PROPER DISTANCE OF A LUMINOUS SOURCE

How do we calculate distances in the expanding universe? We should generalise the definition of *proper distance* discussed in the case of special relativity. This is the distance measured by observers at rest with the ruler to be measured. How can the measurement be performed in practice? An observer at one edge can send a light signal to an observer on the other edge who reflects the signal back. The time delay times the speed of light divided by two would then give the size of the ruler or, equivalently, the distance between the two observers.

We are interested in calculating the proper distance between a luminous source and us but in the expanding universe there is an evident obstacle. While the light signal is travelling, for example between a galaxy located in a point A with radial comoving coordinate r (assumed at rest in the comoving system) and us (conveniently located in the origin O), the space itself between the galaxy and us is expanding $\propto a(t)$. One should be able to make a measurement at a given time in order to remove the biasing effect of the cosmological expansion. In principle, the only way this could be done is by having a chain of an infinite number of neighbouring observers able to exchange at the same time t a light signal with the closest neighbour placed at an infinitesimal distance $d\ell$. The total distance would then be obtained integrating over all the infinitesimal intervals of time that the light takes to travel between two neighbour observers. This is the definition of *proper distance* of a luminous source in the expanding universe extending the definition discussed in special relativity. Mathematically, this translates in calculating the integral

$$d_{\mathrm{pr}}(t, r) \equiv a(t)\, R_0 \int_0^r \frac{dr'}{\sqrt{1 - k\, r'^2}} \,. \tag{6.19}$$

Notice that in this case the scale factor $a(t)$ is outside the integral, since the integral is done at a fixed time t. In particular the proper distance of a luminous source at comoving distance r at the present time is given by

$$d_{\mathrm{pr},0}(r) \equiv R_0 \int_0^r \frac{dr'}{\sqrt{1 - k\, r'^2}} \,, \tag{6.20}$$

so that one obtains the relation

$$d_{\mathrm{pr}}(t) = a(t)\, d_{\mathrm{pr},0} \tag{6.21}$$

(having implied the trivial dependence on r). Comparing this result with Eq. (6.1), one can see that the proper distance calculated at the present time coincides with the *comoving distance* of the galaxy (the latter is independent of time since both the observer and the galaxy are assumed at rest in the comoving system) and that the proper distance $d_{\mathrm{pr}}(t)$ is a physical distance scaling linearly with the scale factor during the expansion.

If we now go back to Eq. (6.14), we can notice that $f(r_{\mathrm{em}}) = d_{\mathrm{pr},0}(r_{\mathrm{em}})/R_0$ and this provides a way to calculate the proper distance of a source in terms of the scale factor. Considering that $a(t) \equiv R(t)/R_0$, we can write

$$d_{\mathrm{pr},0}(r_{\mathrm{em}}) = \int_{t_{\mathrm{em}}}^{t_0} \frac{c\, dt}{a(t)} \,. \tag{6.22}$$

One would need a theoretical model that gives $a(t)$ in order to calculate the proper distance. However, in Chapter 10 we will show that, if the redshift of the object is not too large, the integral on the right-hand side can be evaluated, model independently, in terms of the redshift of the source and the Hubble constant. In this way Hubble's law will be derived as an approximated relation between the luminosity distance and the redshift of the source, valid only at sufficiently small redshifts.

6.3 LEMAITRE'S EQUATION

Differentiating Eq. (6.21) with respect to time, we obtain

$$\dot{d}_{\mathrm{pr}}(t) = \dot{a}(t)\, d_{\mathrm{pr},0} \,. \tag{6.23}$$

From this equation, defining $v_{\mathrm{pr}}(t) \equiv \dot{d}_{\mathrm{pr}}(t)$ and introducing (as we did in the Milne–McCrea model) the *expansion rate* (or *Hubble parameter*) as

$$H(t) \equiv \frac{\dot{a}}{a}, \tag{6.24}$$

we obtain *Lemaitre's equation*[5]

$$\boxed{v_{\mathrm{pr}}(t) = H(t)\, d_{\mathrm{pr}}(t)}. \tag{6.25}$$

This can be specified at the present time obtaining

$$v_{\mathrm{pr}}(t_0) = H_0\, d_{\mathrm{pr},0}, \tag{6.26}$$

where $H_0 \equiv H(t_0)$, as we will see in Chapter 10, will be identified with the *Hubble constant* in Eq. (2.20).

Notice that we derived an analogous expression, Eq. (3.6), also within the Milne–McCrea model, a Newtonian approach to the universe expansion. The difference is that in the Milne–McCrea model the recession velocities of galaxies are due to their propagation in space and in this case one can use the Doppler effect relation to express straightforwardly recession velocities in terms of redshifts. Here, in the case of Friedmann cosmology based on general relativity, the Lemaitre equation is a consequence of the expansion of space itself and the relation between recession (proper) velocity and redshift is less straightforward and will be discussed in Chapter 10 in a model independent way. We will discover that Hubble's law is valid only at redshifts $z \lesssim 1$, while at higher redshifts there are deviations that cannot be reproduced by the Milne–McCrea model and actually not even by the simplest Friedmann cosmological models but will necessarily require the presence of a so-called dark energy component.

In the Milne–McCrea model we had to impose the upper bound Eq. (3.23) on the size of the universe in order to avoid superluminal expansion velocities. Now, in Friedmann cosmological models, there is not such an upper bound since superluminal velocities due to the expansion of the metric do not violate *causality*. This is because a causal signal (e.g., light) exchanged between two receding objects (e.g., two galaxies at rest in the comoving system) would still travel at the speed of light between points at an infinitesimal physical distance in the comoving system. Causality imposes that *physical velocities*, corresponding to the peculiar velocities of objects with respect to coordinate comoving points, are not higher than the speed of light.

---

[5]It was first derived in [2]. It is often confused with Hubble's law when the non-relativistic Doppler effect relation $z = v/c$ is used to replace redshift with velocity. However, we will show in Chapter 10 that Hubble's law follows from Eq. (6.22) in the limit of low redshifts $z \ll 1$ and is, therefore, an approximated expression relating observable quantities while Lemaitre's equation is an exact property of a formal quantity, the proper distance, and is a direct consequence of the cosmological principle. Hubble's law and Lemaitre's equation are, therefore, related but not equivalent. Sometimes Eq. (6.25) is called *idealised or theoretical Hubble's law* but since it is conceptually different from Hubble's law and since it was derived by Lemaitre [2], it is appropriate to name it *Lemaitre's equation*.

6.4 THE TWO FUNDAMENTAL EQUATIONS OF FRIEDMANN COSMOLOGY

The FRW metric is particularly simple. It is described in terms of just three parameters: the curvature parameter k, the curvature radius at present R_0 and the scale factor $a(t)$. However, only the scale factor depends on time (the curvature parameter cannot change during the expansion) and its evolution describes how the gravitational field (the metric tensor) evolves in a homogeneous and isotropic ($\equiv$ Friedmann) universe: in other words all the dynamics of a Friedmann universe is simply described by the scale factor time evolution $a(t)$.

As we have seen, in general relativity the dynamics of the gravitational field, described by the metric, is described by Einstein's equations (4.118). Therefore, we need to specialise these equations to the case of a homogeneous and isotropic universe using the FRW metric.

Since the cosmological principle applies, there cannot be any equation privileging a particular spatial component. This simple observation is confirmed by an explicit calculation of the Einstein tensor $G_{\mu\nu}$ showing that all equations involving spatial indexes ($\mu = i, \nu = j$) produce trivial identities. Therefore, Einstein's equations reduce just to one time-time equation and, since $g_{00} = 1$, one obtains

$$R_{00} - \frac{1}{2}\mathcal{R} = \frac{8\pi G}{c^4} T_{00} . \tag{6.27}$$

It can be shown that

$$R_{00} = -\frac{3}{c^2}\frac{\ddot{a}}{a} , \tag{6.28}$$

and that the Ricci scalar is given by

$$\mathcal{R} = -\frac{6}{c^2}\left[\frac{\ddot{a}}{a} + \frac{\dot{a}^2}{a^2} + \frac{k\,c^2}{a^2\,R_0^2}\right] . \tag{6.29}$$

The energy-momentum tensor of a homogeneous and isotropic universe has necessarily to reduce to the form valid for an ideal fluid (cf. Eq. (4.71)) so that $T^{00} = \varepsilon$.

Plugging these three results into Eq. (6.27), we finally find the *Friedmann equation* [4]

$$\boxed{H^2(t) = \frac{8\pi G}{3\,c^2}\varepsilon(t) - \frac{k\,c^2}{a^2(t)\,R_0^2}} , \tag{6.30}$$

where remember that the Hubble parameter is defined as $H \equiv \dot{a}/a$. Notice that the Friedmann equation is not a *closed* differential equation for $a(t)$, since it contains another unknown time-dependent quantity $\varepsilon(t)$. However, we still have to impose the *energy-momentum tensor conservation equation* Eq. (4.120). We have seen that in special relativity this corresponds to a generalisation of the first principle of thermodynamics (see Eq. (4.76)).

In a Friedmann universe it can be specialised in a very simple way. First of all now the fluid is strictly homogeneous and, therefore, the volume V can be chosen arbitrarily large. Second, the fluid cannot undergo ordinary compression or dilatation since this would result in inhomogeneities that we are neglecting. If this were the whole story, then one would just obtain $d(\varepsilon V) = $ const, conservation of energy.

However, the fluid is now subject to the cosmological expansion (that is not a usual dilatation) and this has to be taken into account. Consider first again the non-relativistic case, where $\varepsilon = \rho c^2$ and the pressure p is negligible. For example, this would correspond to the case where all the energy is stored in the mass of galaxies and their peculiar velocities are negligible. For the sake of simplicity, assume also that all galaxies have the same mass M_G and that their number density is given by n_G. In this case one would simply have $\rho = n_G M_G$. Consider some galaxies are at rest in a portion of comoving volume $V_{(0)} = R_0^3 \Delta r^3$ corresponding to an expanding physical volume $V(t) = V_{(0)} a^3(t)$. If their number is conserved, so that $N_G = n_G(t) R_0^3 \Delta r^3 a^3(t) = N_{G,0}$, the density would then simply get diluted during the expansion as $n_G(t) = n_{G,0} a^{-3}(t)$ (going back to the 2-dim analogy: the density of 'ants' on the surface of an expanding balloon would decrease as an effect of the expansion). Therefore, in this particular case the conservation of the number of galaxies would simply imply $d(n_G a^3)/dt = 0$ or equivalently $d(\varepsilon a^3)/dt = 0$, with $\varepsilon = \rho c^2$.

In a more general situation, where galaxies do have peculiar velocities (implying that pressure now is not completely vanishing) and the energy is not just stored in the mass of galaxies, this result generalises in a very simple way. The first principle of thermodynamics holds as in special relativity, but it applies to the gas contained in a constant portion of *comoving* volume $V_{(0)}$. Therefore, the *energy-momentum tensor conservation equation in the Friedmann universe* would simply read as

$$\frac{d(\varepsilon a^3 V_{(0)})}{dt} = -p \frac{d(a^3 V_{(0)})}{dt}, \tag{6.31}$$

and dividing by the arbitrary constant $V_{(0)}$, one obtains the *cosmological fluid equation*

$$\boxed{\frac{d(\varepsilon a^3)}{dt} = -p \frac{d a^3}{dt} \Leftrightarrow \dot{\varepsilon} = -3 \frac{\dot{a}}{a} (\varepsilon + p)}, \tag{6.32}$$

that we wrote also in a second fully equivalent form that is more convenient to use in general for $p \neq 0$.[6]

---

[6]This intuitive way to understand the cosmological fluid equation is confirmed by a rigorous derivation following from the *energy-momentum tensor conservation equation* in the presence of a gravitational field, Eq. (4.120). If we specialise this equation to the case of the FRW metric, we find that similarly to what happened with Einstein's equations, the spatial components simply give trivial identities and one is left only with the time component that, using

$$\Gamma^0{}_{00} = 0 ; \quad \Gamma^1{}_{01} = \Gamma^2{}_{02} = \Gamma^3{}_{03} = \frac{\dot{a}}{a}, \tag{6.33}$$

gives directly the *fluid equation*

$$\dot{\varepsilon} = -3 \frac{\dot{a}}{a} (\varepsilon + p). \tag{6.34}$$

Let us now explicitly derive the *acceleration equation* for Friedmann cosmological models. It is quite simple if one first of all recasts the Friedmann equation Eq. (6.30) multiplying both sides by a^2, obtaining

$$\dot{a}^2 = \frac{8\,\pi\,G}{3\,c^2}\,\varepsilon\,a^2 - \frac{k\,c^2}{R_0^2}\,.$$

(6.35)

In this way, differentiating this equation with respect to time, the constant curvature term cancels out and, dividing by $2\,\dot{a}$, one can write

$$\ddot{a} = \frac{8\,\pi\,G}{3\,c^2}\,\varepsilon\,a + \frac{8\,\pi\,G}{3\,c^2}\,\frac{a^2\,\dot{\varepsilon}}{2\,\dot{a}}\,,$$

(6.36)

or, equivalently,

$$\ddot{a} = \frac{4\,\pi\,G}{3\,c^2}\left(2\,\varepsilon\,a + a^2\,\frac{\dot{\varepsilon}}{\dot{a}}\right)\,.$$

(6.37)

If we now insert the fluid equation (6.34), recast as

$$\frac{\dot{\varepsilon}}{\dot{a}} = -3\,\frac{p+\varepsilon}{a}\,,$$

(6.38)

we first obtain

$$\ddot{a} = \frac{4\,\pi\,G}{3\,c^2}\left[2\,\varepsilon\,a - 3\,(p+\varepsilon)\,a\right]\,,$$

(6.39)

and then, finally,

$$\boxed{\ddot{a} = -\frac{4\,\pi\,G}{3\,c^2}\,(\varepsilon + 3p)\,a}\,.$$

(6.40)

As already anticipated in Chapter 3, this equation represents a clear departure from Newton's gravitation law, since it clearly shows that not only the mass determines the gravitational acceleration but even the kinetic energy, contributing both to the total energy density ε and to the pressure p.

6.5 CRITICAL ENERGY DENSITY

Defining the *critical energy density* $\varepsilon_{\mathrm{c}}(t)$ as the total energy density in a flat universe, one immediately finds

$$\boxed{\varepsilon_{\mathrm{c}}(t) \equiv \frac{3\,H^2(t)\,c^2}{8\,\pi\,G}}\,.$$

(6.41)

At the present time one has

$$\varepsilon_{\mathrm{c},0} \equiv \frac{3\,H_0^2\,c^2}{8\,\pi\,G} \simeq 10.54\,h^2\,\mathrm{GeV\,m^{-3}}\,,$$

(6.42)

having used the Eqs. (2.10), (2.1), (2.9) and (2.28) (*see problem n. 6*).[7] The Friedmann equation (6.30) can then be recast as

$$H^2(t) = H^2(t) \frac{\varepsilon(t)}{\varepsilon_c(t)} - \frac{k\,c^2}{a^2(t)\,R_0^2} \, . \tag{6.43}$$

If we further introduce the *energy density parameter*,

$$\boxed{\Omega(t) \equiv \frac{\varepsilon(t)}{\varepsilon_c(t)}}, \tag{6.44}$$

we can write

$$H^2(t) = H^2(t)\,\Omega(t) - \frac{k\,c^2}{a^2(t)\,R_0^2} \, . \tag{6.45}$$

Specialising this equation at the present time,

$$H_0^2 = H_0^2\,\Omega_0 - \frac{k\,c^2}{R_0^2} \, , \tag{6.46}$$

one finally finds an expression of the curvature term in terms of the observables H_0 and Ω_0,

$$\frac{k\,c^2}{R_0^2} = -H_0^2\,(1 - \Omega_0) \, . \tag{6.47}$$

This expression shows that the curvature parameter k is determined just by Ω_0, explicitly:

- $\Omega_0 < 1 \;\Leftrightarrow\; k = -1 \;\Leftrightarrow\;$ open universe;

- $\Omega_0 = 1 \;\Leftrightarrow\; k = 0 \;\;\;\Leftrightarrow\;$ flat universe;

- $\Omega_0 > 1 \;\Leftrightarrow\; k = +1 \;\Leftrightarrow\;$ closed universe.

If we now go back to Eq. (6.43), we can express the curvature term using Eq. (6.47) and, considering that

$$\frac{H^2}{\varepsilon_c(t)} = \frac{H_0^2}{\varepsilon_{c,0}} = \frac{8\,\pi\,G}{3\,c^2} \, , \tag{6.48}$$

we can first write

$$H^2(t) \equiv \left(\frac{\dot{a}}{a}\right)^2 = H_0^2\,\frac{\varepsilon(t)}{\varepsilon_{c,0}} + \frac{H_0^2\,(1 - \Omega_0)}{a^2} \tag{6.49}$$

and then finally

$$\boxed{\dot{a}^2(t) = H_0^2\,\Omega_0\,a^2(t)\,\frac{\varepsilon(t)}{\varepsilon_0} + H_0^2\,(1 - \Omega_0) \, .} \tag{6.50}$$

---

[7]The value of h can be left unspecified since in the end one mainly needs the (precisely known) value of $\varepsilon_{c,0}\,h^{-2}$, rather than the less precise value of $\varepsilon_{c,0}$ affected by the error on h^2.

This is the Friedmann equation expressed in terms of the Hubble constant H_0 and the total energy density parameter at present Ω_0.

In general we can say that for $\Omega_0 \leq 1$ the expansion of the universe will continue forever. However, we cannot say in general whether it will always decelerate or not, since this depends on whether $\varepsilon\, a^2$ monotonically decreases with time or not or, equivalently, looking at the acceleration equation Eq. (6.39), whether $\varepsilon + 3p$ is always positive or turns negative at some stage.

Moreover, for the case $\Omega_0 > 1$, nothing can be said without knowing how $\varepsilon\, a^2$ behaves, since now the two terms have opposite sign and there could be potentially a special time where the expansion stops turning into a collapse.

6.6 FRIEDMANN EQUATION AS A CONSERVATIVE SYSTEM

We can further express the Friedmann equation (6.50) writing

$$\frac{\dot{a}^2(t)}{H_0^2} = \Omega_0\, a^2(t)\, \frac{\varepsilon(t)}{\varepsilon_0} + (1 - \Omega_0)\,. \tag{6.51}$$

This dimensionless form is particularly suitable for some considerations that will prove useful in the next chapter when we will discuss the solutions of the Friedmann equation within so-called Lemaitre models.

From a mathematical point of view the Friedmann equation is a *conservative system*, since there is a clear conserved quantity, mathematically an *integral of motion*, that we can call *effective energy of the system*, defined by

$$E(a) \equiv \frac{\dot{a}^2(t)}{H_0^2} - a^2\, \Omega_0\, \frac{\varepsilon(a)}{\varepsilon_0}\,, \tag{6.52}$$

that during the expansion is constantly equal to[8]

$$E_0 \equiv 1 - \Omega_0\,. \tag{6.53}$$

This constant of motion should be regarded as an effective (dimensionless) total energy of the system. We have seen that in the Newtonian interpretation, this property of the equation was somehow built-in since one started indeed from energy conservation to derive it. In this approach the effective energy was indeed proportional to the the total energy of a galaxy ejected apart from the centre of the explosion (more precisely we found $E_0 = C_0/H_0^2$). This time E_0 is not related to some actual kind of energy, rather to the curvature of the universe, and it should be, therefore, regarded as an effective energy.

Analogously we can introduce an (dimensionless) *effective potential* defined as

$$V(a) \equiv -a^2\, \Omega_0\, \frac{\varepsilon(a)}{\varepsilon_0}\,, \tag{6.54}$$

---

[8]In the literature this constant is often indicated with $\Omega_{k,0}$.

so that the dimensionless Friedmann equation can now be explicitly expressed as a one-dimensional conservative system,

$$E(a) \equiv \frac{\dot{a}^2(t)}{H_0^2} + V(a) = E_0 , \tag{6.55}$$

where the only degree of freedom is the scale factor $a(t)$. This form shows that no matter how complicated the scale factor time evolution is, the quantity on the left-hand side, the integral of the system, remains constant during the motion. This mathematical property of the Friedmann equation will prove very useful in order to understand its solutions in a general way.

EXERCISES

Exercise 6.1 *Consider a freely propagating proton with energy $E = 2.814\,\text{GeV}$ at a (past) time t such that $a(t) = 1/2$. Using $m_p = 0.938\,\text{GeV}/c^2$, what is its energy at the present time?*

Exercise 6.2 *The proper distance between two far galaxies having exactly the same cosmological redshift $z = 1$ is $10\,\text{Mpc}$ at present. What was the value of the proper distance at the time emission?*

Exercise 6.3 *Derive the fluid equation combining the Friedmann equation with the acceleration equation.*

Exercise 6.4 *Derive the Friedmann equation combining the fluid equation with the acceleration equation.*

Exercise 6.5 *Show that the expression*

$$\varepsilon_c(t) \equiv \frac{3\,H^2(t)\,c^2}{8\,\pi\,G} , \tag{6.56}$$

for the critical energy density is dimensionally correct.

Exercise 6.6 *Find the numerical value for the critical energy density at present $\varepsilon_{c,0}$ in $\text{GeV}\,\text{m}^{-3}$, expressing it in terms of the parameter h defined as $h \equiv H_0/(100\,\text{km}\,\text{s}^{-1}\,\text{Mpc}^{-1})$ and using $M_P = 1.2 \times 10^{19}\,\text{GeV}/c^2$.*

Exercise 6.7 *We expressed the Friedmann equation in the useful form (6.50), in terms of H_0 and Ω_0. However, this is not always valid. In which case does one need to use the original form Eq. (6.30)?*

BIBLIOGRAPHY

[1] E. K. Conklin, *Velocity of the Earth with respect to the cosmic background radiation*, Nature, **222** (1969) 971.

[2] G. Lemaitre, Annales Soc. Sci. Brux. Ser. I Sci. Math. Astron. Phys. A **47** (1927) 49.

[3] A. Friedman, *On the curvature of space*, Z. Phys. **10** (1922) 377 [Gen. Rel. Grav. **31** (1999) 1991].

CHAPTER 7

Building a cosmological model

I n this chapter we will first make some general considerations on the steps that are needed for building a cosmological model. We will start considering simple cosmological models containing ordinary fluids like matter and radiation drawing a general strategy to build a cosmological model and derive all its properties.

7.1 GENERAL STRATEGY

How to *build* a cosmological model? One needs some additional condition on $\varepsilon(t)$ to be plugged into the Friedmann equation (6.30) in order to obtain a closed differential equation. This clearly has to be done exploiting the fluid equation (6.32). However, this usually also introduces an additional unknown quantity, pressure, into the set of equations. It is then necessary to specify an *equation of state* of the form $p = p(\varepsilon)$[1] in order to link pressure p and energy density ε. The equation of state corresponds to specifying the nature of the fluid (or fluids) filling the universe and is, therefore, a very important link between cosmology and fundamental physics. This means that general relativity alone is not sufficient to specify the cosmological model.

The general strategy for building a cosmological model can, therefore, be summarised in three steps:

- The first step consists in making an assumption on the nature of the fluid, specifying its *equation of state* $p = p(\varepsilon)$. This of course includes the possibility of an admixture of different components.

- The second step consists in replacing the equation of state into the fluid equation, so that one now obtains a closed differential equation

$$\frac{d(\varepsilon\, a^3)}{dt} = -p(\varepsilon)\, \frac{d\, a^3}{dt}\,, \tag{7.1}$$

from which one obtains a relation

$$\varepsilon = \varepsilon(a)\,, \tag{7.2}$$

---

[1]In the case of inflation, discussed in Chapter 15, there are fluids whose equation of state cannot be reduced to the simple form $p = p(\varepsilon)$ and require a more sophisticated description.

since time cancels out. This relation specifies the energy density dependence on the scale factor during the expansion.

- Plugging this relation into the Friedmann equation (6.50), one finally obtains a closed differential equation for the scale factor a, from which one obtains a solution for the time evolution of the scale factor $a(t)$. This can then be plugged back into Eq. (7.2) obtaining an explicit relation $\varepsilon(t)$.

It is quite remarkable that, starting from an assumption on the nature of the fluid specifying the equation of state, one can describe all the details of the cosmological expansion by means just of the Friedmann equation and the fluid equation. We now discuss a few simple remarkable cases that will provide concrete examples of this general strategy.

7.2 THE SIMPLEST MODEL: THE EMPTY UNIVERSE (MILNE MODEL)

The simplest cosmological model one can imagine is the *empty universe* [1], relying on the simple, apparently trivial, assumption $\varepsilon = 0$. This of course implies necessarily $\Omega_0 = 0$ so that Eq. (6.50) reduces simply to $\dot{a}^2 = H_0^2$. Considering only the expanding solution ($\dot{a} > 0$), as imposed by the observations, one immediately obtains

$$\dot{a} = H_0 , \tag{7.3}$$

whose solution is simply given by

$$a(t) = H_0 t + a(0) , \tag{7.4}$$

where the constant $a(0) \equiv a(t = 0)$ has to be determined by the initial conditions. Notice that this solution implies the existence of a special time t_s such that $a(t_s) = 0$ and this is given by

$$t_s = -a(0) H_0^{-1} , \tag{7.5}$$

in a way that one can write $a(t) = H_0 (t - t_s)$. This special time corresponds to the so-called *initial singularity*, since the FRW metric is physically meaningless at this time. The existence of the initial singularity signals that going back in time, the validity of Friedmann cosmology breaks down at some point, before reaching the singularity and it cannot give any description on what happens too close to the singularity or before. It is, therefore, quite reasonable to choose *conventionally* t_s as the *origin of time* setting $t_s = 0$. In this way $a(0) = 0$ and time cannot be negative. With this choice the *age of the universe*, conventionally defined as the time elapsed since the initial singularity, is simply given by the *cosmic time t*.

The existence of the initial singularity is not specific to the empty model but, as we will see, it is actually a generic feature of a wide class of models, so-called *Big Bang models*,[2] including that one currently supported by all observations: it is one of the biggest problems of modern cosmology.

---

[2]As should be clear at this point, the name Big Bang should not be interpreted literally, as if

In the empty universe the (current) age of the universe t_0, corresponding to $a_0 \equiv a(t_0) = 1$, is simply given by the *Hubble time* (see Eq. (2.30)), namely

$$t_0 = H_0^{-1} = (14.4 \pm 0.2)\,\mathrm{Gyr}\,, \tag{7.6}$$

where we used the latest *Planck* satellite H_0 determination Eq. (2.26). A strictly empty universe does not describe any physically interesting situation. It can be even proven that it is actually equivalent to a spatially infinite static universe and in this case the notion of *age of the universe* would lose meaning.

However, the empty universe should be more interestingly regarded as the asymptotic limit for $\Omega_0 \to 0$ of any model with generic equation of state $p = p(\varepsilon)$. We will see a few examples showing explicitly this interesting property.

7.3 MATTER UNIVERSE

We have already considered the case when all matter and energy in the universe is described by a fluid made of massive objects (e.g., galaxies for which the stellar motion is negligible) at rest in the comoving system, i.e., with negligible peculiar velocities. In cosmology this kind of fluid is usually referred to as *matter*[3] and we will denote it with the symbol M. The energy density is dominated by the mass contribution so that $\varepsilon_M \simeq \rho\,c^2$. Matter is pressureless so that the equation of state is simply given by

$$p_M = 0\,. \tag{7.7}$$

This can be understood considering that pressure is the effect of the kinetic motion of the fluid in the comoving system.[4] With such a simple equation of state, the fluid equation in the form given by Eq. (6.32), immediately yields the result $\varepsilon_M\,a^3 = $ const, or in terms of quantities at present

$$\varepsilon_M(a) = \frac{\varepsilon_{M,0}}{a^3} \propto a^{-3}\,. \tag{7.9}$$

This confirms the result derived already in Section 6.4 specifically for matter, using the conservation of mass in a comoving volume. It was the second step of the general strategy for cosmological model building applied to the specific case of matter.

---

the universe had originated from some privileged point of space as in the Milne–McCrea model. Rather it should be interpreted as the universe initial state corresponded to a value of the scale factor that, as we will discuss, had to be at least as small as 10^{-10} in order to explain successfully the primordial nuclear abundances. However, we will see that the theory of inflation suggests that it had to be actually even many orders of magnitude smaller than 10^{-10}.

[3]Sometimes also *dust* is used but we will always use the word matter.

[4]It can be shown that for a gas with a generic distribution function $f(|\vec{p}|)$ (the mean occupation number of each quantum state) the pressure, written in the comoving system, is given by (natural system)

$$p(t) = \int \frac{d^3 p}{(2\pi)^3}\, \frac{|\vec{p}|^2}{3\,E}\, f(|\vec{p}|, t)\,. \tag{7.8}$$

The third and last step consists in plugging the result found for $\varepsilon(a)$ in the Friedmann equation (6.50), obtaining easily

$$\dot{a}^2(t) = H_0^2\,\Omega_0\,a^{-1}(t) + H_0^2\,(1 - \Omega_0)\,. \tag{7.10}$$

In addition, we can also specialise the acceleration equation (6.40),

$$\ddot{a} = -\frac{4\,\pi\,G}{3\,c^2}\,\frac{\varepsilon_{M,0}}{a^2}\,, \tag{7.11}$$

showing that the expansion always decelerates.

We can now specialise some of the general considerations discussed after Eq. (6.50) about the fate of the expansion. The analysis proceeds practically along the same lines discussed in Chapter 3, when we derived the Friedmann equation within a Newtonian approach. However, now the physical interpretation of the expansion is quite different, as we extensively pointed out. We need to distinguish three cases.

(i) For $\Omega_0 < 1$ (open space) the expansion lasts forever and the asymptotic expansion velocity is given by $\dot{a}(t = \infty) = H_0\,\sqrt{1 - \Omega_0}$ [2].

(ii) For $\Omega_0 = 1$ (flat space) the expansion lasts forever as well but the asymptotic velocity vanishes ($\dot{a}(t = \infty) = 0$). Moreover in this case one can find easily an explicit analytical solution. The Eq. (7.10) specialises into

$$\sqrt{a}\,\dot{a}(t) = H_0 \quad \Rightarrow \quad \frac{da^{3/2}}{dt} = \frac{3}{2}\,H_0 \tag{7.12}$$

and setting again, as we did in the case of the empty universe, the origin of time at the singularity ($a(t = 0) = 0$), one immediately finds the solution

$$a(t) = \left(\frac{t}{t_0}\right)^{\frac{2}{3}}\,, \tag{7.13}$$

where the age of the universe t_0 is given by

$$t_0 = \frac{2}{3}\,H_0^{-1} = (9.60 \pm 0.15)\,\text{Gyr}\,. \tag{7.14}$$

A flat matter model (*Einstein-de Sitter model* [3]) would certainly be an attractive simple model but this result for the age of the universe is in contradiction with the age of the oldest stars in globular clusters that are certainly older than 10 Gyr [7, 8]. Therefore, there would be no enough time for these stars to form in a flat matter-dominated universe. We will discuss in Chapter 9 the possible ways out of this *age problem* and in Chapter 10 we will be able to pin down a specific one currently supported by the observations. Finally, notice that from the solution found for $a(t)$ one can deduce the following behaviour for $\varepsilon_M(t)$

$$\varepsilon_M(t) = \varepsilon_{M,0}\left(\frac{t_0}{t}\right)^2\,. \tag{7.15}$$

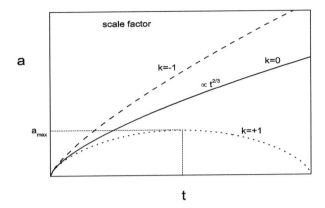

FIGURE 7.1: Scale factor evolution in the matter-dominated model for the three possible values of the curvature parameter k.

(iii) For $\Omega_0 > 1$ (closed space) there is a special time when the expansion stops and turns into a collapse. [5] This corresponds to a maximum value of the scale factor given by

$$a_{\max} = \frac{\Omega_0}{\Omega_0 - 1}.$$
(7.16)

The results for the universe expansion in a matter universe are summarised in Fig. 7.1.

7.4 RADIATION UNIVERSE

Let us now assume that the universe is dominated by radiation denoted with R. In this case the equation of state is given by[6]

$$p_R = \frac{1}{3}\,\varepsilon_R.$$
(7.18)

---

[5] A closed matter model is also called the *Friedmann–Einstein model* [4, 5]. It is interesting since it is the first model Einstein considered after the 1917 static model ('his biggest blunder'), where he embraces the idea of an expanding universe proposed by A. Friedmann. Curiously it is also the subject of his 2nd Rhodes lecture that he gave at Oxford University in 1931. One of the blackboards used by Einstein (*Einstein's blackboard*) has never been erased and is preserved at the *Museum of the History of Science* in Oxford. Recently, it has been noticed that the original Einstein's paper contains a numerical mistake, even visible on the blackboard itself [6].

[6] This is a well-known result in classical electromagnetism that is confirmed also by treating radiation as a gas of photons. If in Eq. (7.8) one takes $|\vec{p}| = E$, then it follows immediately considering that (natural system)

$$\varepsilon = \int \frac{d^3 p}{(2\,\pi)^3}\, E\, f(|\vec{p}|, t).$$
(7.17)

If this equation of state is inserted into the second equivalent form of the fluid equation in Eq. (6.32), one has

$$\frac{\dot{\varepsilon}_R}{\varepsilon_R} = -4\frac{\dot{a}}{a}, \tag{7.19}$$

that immediately yields

$$\varepsilon_R = \frac{\varepsilon_{R,0}}{a^4} \propto a^{-4}. \tag{7.20}$$

How can we physically understand this result? In the case of matter we have seen that simply $\varepsilon_M = \rho c^2 = n_G \overline{M}_G c^2 \propto a^{-3}$.

In the case of radiation, supposing this is made of photons with mean energy $\overline{E}_\gamma$, it can be easily understood as a consequence of the momentum redshift, since $\overline{E}_\gamma = c\overline{|\vec{p}|} \propto a^{-1}$. Therefore, now $\varepsilon_R = n_\gamma \overline{E}_\gamma \propto a^{-4}$.[7] Notice that for radiation in thermal equilibrium, since one has $\varepsilon_R \propto T^4$ (see Eq. (2.36)), one immediately finds the result

$$T \propto a^{-1}. \tag{7.22}$$

This result can be regarded as the macroscopic consequence of the momentum redshift.[8] In the case of radiation the Friedmann equation (6.50) specialises into

$$\dot{a}^2(t) = H_0^2\,\Omega_0\,a^{-2}(t) + H_0^2\,(1 - \Omega_0). \tag{7.25}$$

Analogously to the case of the Einstein–de Sitter model, one has to distinguish three cases:

(i) For the case of an open universe, $\Omega_0 < 1$, the expansion lasts forever and the asymptotic expansion velocity is given by $\dot{a}(t = \infty) = H_0\sqrt{1 - \Omega_0}$.

(ii) For the case of a closed universe, $\Omega_0 > 1$, there will be a time, corresponding to $a = a_{max} \equiv \sqrt{\Omega_0/(\Omega_0 - 1)}$, when the expansion stops and turns into a collapse.

---

[7]This conclusion holds, even more generally, considering that (natural system)

$$\varepsilon_R = \varepsilon_\gamma = g_\gamma \int_0^\infty \frac{d|\vec{p}|}{2\pi^2\,\hbar^3}\,|\vec{p}|^3\,f(\vec{p}, t) \propto a^{-4}. \tag{7.21}$$

The last passage holds because we are assuming that photons either are non-interacting or in thermal equilibrium and, therefore, $f(\vec{p}, t) = $ const.

[8]For a gas of photons in thermal equilibrium one has (natural system)

$$\varepsilon_R = \varepsilon_\gamma = g_\gamma \int_0^\infty \frac{d|\vec{p}|}{2\pi^2}\,|\vec{p}|^3\,f_{BE}(\vec{p}, t), \tag{7.23}$$

where

$$f_{BE}(\vec{p}, t) = \frac{1}{e^{|\vec{p}|/T} - 1}. \tag{7.24}$$

If photons are not interacting or in thermal equilibrium, as we are assuming, then $f_{BE}(\vec{p}, t) = $ const (i.e., $df/dt = 0$ where notice that we are taking the total derivative with respect to time). Since $|\vec{p}| \propto a^{-1}$ then necessarily $T \propto a^{-1}$: everything is consistent.

(iii) For a *flat universe* ($\Omega_0 = 1$), the expansion again lasts forever but the asymptotic velocity vanishes. However, this time one has a different result for the scale factor dependence on time. In this case the Friedmann equation specialises into

$$\dot{a}^2(t) = H_0^2\, a^{-2}(t)\,. \tag{7.26}$$

It is then straightforward to see that from this equation one finds first $a^2(t) = 2\,H_0\,t$ and then

$$a(t) = \left(\frac{t}{t_0}\right)^{1/2}, \tag{7.27}$$

where now the age of the universe is given by

$$t_0 = \frac{1}{2}\, H_0^{-1} = (7.05 \pm 0.12)\,\text{Gyr}. \tag{7.28}$$

It is clear that in a pure radiation model the *age problem* gets exacerbated compared to the matter model. We will see, however, that the radiation model is important since it describes a significant regime holding in the early stages of the history of the universe. Notice that again, as in the matter universe, one has

$$\varepsilon_R(t) = \varepsilon_{R,0} \left(\frac{t_0}{t}\right)^2. \tag{7.29}$$

7.5 ONE-FLUID MODELS WITH $p/\varepsilon = w = \text{const}$

Both matter and radiation models can be regarded as two particular cases of a general class of one-fluid models for which the equation of state can be written in the form

$$p = w\,\varepsilon\,, \tag{7.30}$$

where $w = \text{const}$. In the case of matter $w_M = 0$ and in the case of radiation $w_R = 1/3$. These are the two standard cases of fluids but as we will see one can envisage more unusual situations, especially considering that we are testing physics on cosmological scales and, therefore, non-standard situations should not be excluded on the basis of ground-based experiments. This is a first example of how the whole universe can be used as a laboratory for fundamental physics. We will see that this open-minded attitude will lead to important conclusions for *particle physics* relying on cosmological considerations and observations.

Let us proceed along the same lines discussed for matter and radiation. The fluid equation, in the second equivalent form in Eq. (6.32), can be recast as

$$\frac{\dot{\varepsilon}}{\varepsilon} = -3\,(1+w)\,\frac{\dot{a}}{a}\,, \tag{7.31}$$

yielding immediately the solution

$$\varepsilon(a) = \frac{\varepsilon_0}{a^{3\,(1+w)}}\,. \tag{7.32}$$

If we limit our analysis to the case of a flat universe ($\Omega_0 = 1$) and if we select the expanding solution in the Friedmann equation Eq. (6.50), we can write

$$\dot{a}(t) \, a^{\frac{1+3\,w}{2}} = H_0 \, . \tag{7.33}$$

Barring the special case $w = -1$, that will require some special consideration, we can write

$$\frac{da^{\frac{3\,(1+w)}{2}}}{dt} = \frac{3\,(1+w)}{2} \, H_0 \, . \tag{7.34}$$

If we further restrict the analysis to models with a singularity, then one has to impose $3\,(1+w)/2 > 0$, equivalent to $w > -1$.[9] In this case, defining once again the origin of time such that $a(t = 0) = 0$, one finds for $w > -1$

$$a(t) = \left(\frac{t}{t_0} \right)^{\frac{2}{3\,(1+w)}} \, , \tag{7.35}$$

with the age of the universe given by

$$t_0 = \frac{2}{3\,(1+w)} \, H_0^{-1} \, . \tag{7.36}$$

One can easily verify that for $w = 0$ and $w = 1/3$ the results obtained respectively for the matter and radiation universe are recovered.

It is interesting to notice that for $w < 0$ one has $t_0 > 2\,H_0^{-1}/3 \simeq 9\,\mathrm{Gyr}$ with t_0 increasing monotonically and going to infinite for $w \to -1$. Therefore, there exist values $-1 < w < 0$ such that one can reconcile t_0 with the age of the oldest star given by $t_\star \simeq (14.46 \pm 0.31)\,\mathrm{Gyr}$ [8]. For example for $w = -1/3$ one would obtain $t_0 \simeq H_0^{-1} \simeq 14.4\,\mathrm{Gyr}$, in perfect agreement with $t_\star$. This represents a first possible way out to the *age problem*, though it poses the issue about the physical nature and justification of a fluid with $w \sim -1/3$.

Notice that the behaviour Eq. (7.15) for $\varepsilon(t)$ found for matter and radiation (in the flat case), holds in this more general case as well, not by chance (see Exercise 7.7).

A fluid with $w > 1$ is called *stiff matter*. It is interesting to notice that the so-called *super-stiff limit*, for $w \to \infty$, would correspond to a case where the fluid is so rigid that universe expansion stops, since one has $\dot{a} \to 0 \Rightarrow a(t) \to \mathrm{const}$. We will return to this limit when we introduce the important concept of *horizon* in Section 10.1. The special case $w = -1$, corresponds to the case of a *vacuum energy* fluid and for a flat universe one obtains the *de Sitter model* that we will discuss in detail in the next chapter.

---

[9]The case of models with $w < -1$ will be discussed in Exercise 7.9.

7.6   MULTI-FLUID MODELS AND THE MATTER-RADIATION MODEL

Let us now consider an admixture of different fluids in a way that the total energy density and the total pressure can be written as a sum,

$$\varepsilon = \sum_i \varepsilon_i \,, \qquad p = \sum_i p_i \,, \tag{7.37}$$

of contributions from fluids with different equations of state $p_i = p_i(\varepsilon_i)$.

In this situation, because of the linearity of the fluid equation and assuming that there is no exchange of energy among the different fluids, the solution $\varepsilon(a)$ will be clearly the sum of the solutions of the one-fluid equations, explicitly:

$$\frac{\dot{\varepsilon}_i}{\varepsilon_i + p_i} = -3\frac{\dot{a}}{a} \;\Rightarrow\; \varepsilon_i = \varepsilon_i(a) \;\Rightarrow\; \varepsilon(a) = \sum_i \varepsilon_i(a)\,. \tag{7.38}$$

In particular it is interesting to consider an admixture of fluids with equations of state of the kind $p_i = w_i\,\varepsilon_i$, with $w_i = \text{const}$. Generalising the result obtained in the previous section for a single fluid, we can straightforwardly write the solution as

$$\varepsilon(a) = \sum_i \frac{\varepsilon_{i,0}}{a^{3\,(1+w_i)}}\,. \tag{7.39}$$

A particularly significant example is given by the *matter-radiation model* [9], with an admixture of just two fluids: matter and radiation. This clearly represents the simplest extension of one-fluid models. In this case one simply has

$$\varepsilon(a) = \frac{\varepsilon_{R,0}}{a^4} + \frac{\varepsilon_{M,0}}{a^3}\,. \tag{7.40}$$

Consequently the energy density parameter at present is given by the sum of two contributions

$$\Omega_0 = \Omega_{R,0} + \Omega_{M,0}\,, \tag{7.41}$$

where $\Omega_{M,0} \equiv \varepsilon_{M,0}/\varepsilon_{c,0}$ and $\Omega_{R,0} \equiv \varepsilon_{R,0}/\varepsilon_{c,0}$.

The *matter-radiation equality time* t_{eq} is defined by the condition

$$\varepsilon_R(a_{\text{eq}}) = \varepsilon_M(a_{\text{eq}})\,, \tag{7.42}$$

where $a_{\text{eq}} \equiv a(t_{\text{eq}})$. From equation (7.40) it immediately follows

$$a_{\text{eq}} = \frac{\varepsilon_{R,0}}{\varepsilon_{M,0}} = \frac{\Omega_{R,0}}{\Omega_{M,0}}\,. \tag{7.43}$$

Correspondingly the redshift at the matter-radiation equality is given by

$$z_{\text{eq}} \equiv \frac{1}{a_{\text{eq}}} - 1 = \frac{\Omega_{M,0}}{\Omega_{R,0}} - 1\,. \tag{7.44}$$

For $a \ll a_{\text{eq}}$ one has a *radiation-dominated regime*, the radiation model is recovered

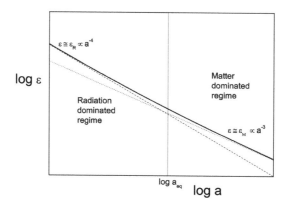

FIGURE 7.2: *Matter-radiation model.* The solid line is the dependence of the energy density ε on the scale factor a (notice the log-log plot). The dashed line is the asymptotic behaviour in the radiation-dominated regime given by $\varepsilon \propto a^{-4}$ while the dotted line is the asymptotic behaviour in the matter-dominated regime, $\varepsilon \propto a^{-3}$.

as an asymptotic limit and approximately

$$a(t) \simeq a_{\mathrm{RD}}(t) \equiv a_{\mathrm{eq}} \left(\frac{t}{t_{\mathrm{eq}}} \right)^{\frac{1}{2}} . \tag{7.45}$$

On the other hand, for $a \gg a_{\mathrm{eq}}$, one has a *matter-dominated regime*, the matter model is recovered as an asymptotic limit and approximately

$$a(t) \simeq a_{\mathrm{MD}}(t) \equiv a_{\mathrm{eq}} \left(\frac{t}{t_{\mathrm{eq}}} \right)^{\frac{2}{3}} . \tag{7.46}$$

For $a \simeq a_{\mathrm{eq}}$ one has a smooth transition between the two regimes (see Exercise 7.5). These results are depicted in Fig. 7.2.

It is instructive to apply the formalism of conservative systems discussed in Section 6.6 to the matter-radiation model. In this case the general effective potential $V(a)$ given by Eq. (6.54) specialises into

$$V_{RM}(a) \equiv -a^2 \left(\frac{\Omega_{R,0}}{a^4} + \frac{\Omega_{M,0}}{a^3} \right) , \tag{7.47}$$

with asymptotic limits

$$\lim_{a \to 0} V_{RM}(a) = -\infty , \qquad \lim_{a \to \infty} V_{RM}(a) = 0 , \tag{7.48}$$

as also confirmed by the plot in Fig. 7.3. Here the possible kinds of dynamical models, solutions of the equation with only matter and radiation, are represented in a graph where on the y axis the effective potential $V(a)$ is plotted versus the

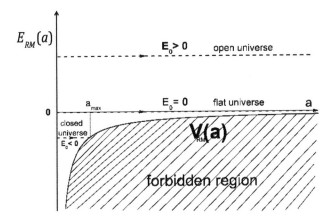

FIGURE 7.3: Classification of matter-radiation models. The expansion at present corresponds to trajectories for increasing a along the lines with constant $E_{RM}(a) = E_0 \equiv (1 - \Omega_0)$. The trajectories cannot access the forbidden region below the effective potential $V_{RM}(a)$. This implies that for a closed universe, when $E_0 = V_{RM}(a)$, the expansion velocity vanishes and the expansion turns into a collapse.

scale factor a. The expansion of the universe corresponds to consider trajectories for increasing a and satisfying

$$E_{RM}(a) \equiv \frac{\dot{a}^2}{H_0^2} + V_{RM}(a) = E_0 \equiv (1 - \Omega_0). \qquad (7.49)$$

The trajectories can never enter into the unphysical region below the effective potential $V_{RM}(a)$, since in this *forbidden region* one would have the unphysical result $E_0 - V(a) = \dot{a}^2/H_0^2 < 0$. For a closed universe ($E_0 < 0$), we have seen that there is a special time when $a = a_{max}$ and a collapsing phase follows afterwards. In this graphical description this corresponds to the time when the trajectory hits the forbidden region ($E_{RM}(a) = V_{RM}(a)$) and the velocity of the expansion vanishes turning negative afterwards. In this way we revisited in a graphical way the results previously obtained analytically for matter-radiation models.

In a general case, considering an admixture of many types of fluids with different values of w_i, the universe undergoes a sequence of consecutive regimes dominated by one-single fluid, necessarily starting, at sufficiently low values of a, from a regime dominated by the fluid with the highest value of w_i, w_{max}, and necessarily ending, at sufficiently high values of a, with a regime dominated by the fluid with the lowest value of w_i, w_{min}. The separation between two consecutive regimes is given by the time of equality between the energy density contribution of the two fluids. It should be noticed, however, that there is not necessarily a dominated regime for each type of fluid, since it could well be that a certain fluid, with $w_{min} < w_i < w_{max}$, never dominates; this is something depending on the values of the $\varepsilon_{i,0}$'s or, equivalently, of the $\Omega_{i,0}$'s. In the next chapter we will discuss in detail the important class of

Lemaitre models, with an admixture of three fluids, that will perfectly illustrate this point.

EXERCISES

Exercise 7.1 *Is the empty model corresponding to an open, flat or closed universe?*

Exercise 7.2 *Knowing that the energy density of starlight radiation is negligible compared to the energy density of the CMB, calculate the value of $\Omega_{R,0} h^2$ using $T = 2.725\,K$ and the value of $\varepsilon_{c,0}$ in Eq. (6.42).*

Exercise 7.3 *Assuming that all matter is in the form of galaxies and using as a number density for galaxies $n_G = 1\,\mathrm{Mpc}^{-3}$ and as a mean galaxy mass $M_G = 10^{12}\,M_\odot$, calculate $\Omega_{M,0} h^2$.*

Exercise 7.4 *What are the values of a_{eq} and z_{eq} using the results of the previous problems? Give an estimate of t_{eq} (in years) in terms of h.*

Exercise 7.5 *Consider the matter-radiation model. Derive an exact solution for the evolution of the scale factor with time that interpolates between the radiation and matter-dominated regime behaviours, respectively $a_{RD}(t)$ and $a_{MD}(t)$.*

Exercise 7.6 *Consider one-fluid cosmological models with $p = w\,\varepsilon$ and $w = \mathrm{const}$. Find the condition on w in order to have an accelerated expansion.*

Exercise 7.7 *We found that for flat models with $p/\varepsilon = w > -1$ energy density scales $\varepsilon(t) \propto t^{-2}$ independently of w. This result has been derived in three steps: first finding a solution $\varepsilon(a)$ from the fluid equation, then finding a solution $a(t)$ from the Friedmann equation and then finally plugging back $a(t)$ in $\varepsilon(a)$. Show that it can be derived directly from the Friedmann equation under proper assumption.*

Exercise 7.8 *Consider a one-fluid model with an equation of state $p = w\,\varepsilon$ and $w = -5/3$. Find the solution $\varepsilon = \varepsilon(a)$; Is there a time when $a(t) = 0$, i.e., does the model have a singularity? Consider the flat case and define $t_0 = H_0^{-1}$; What is the value $a(t = 0)$? What is the value $a(t = 2\,t_0)$? What happens at $t > 2\,t_0$? Finally, give an analytical expression for $a(t)$ and plot $a(t)$.*

Exercise 7.9 *Generalise the results obtained in the previous exercise for a generic value $w < -1$, finding in particular an expression for the time of the 'Big Rip' from now.*

BIBLIOGRAPHY

[1] E.A. Milne, *Relativity gravitation and world structure*, Oxford University Press (1935).

[2] A. Friedmann, *On the possibility of a world with constant negative curvature of space*, Z. Phys. **21** (1924) 326 [Gen. Rel. Grav. **31** (1999) 2001].

[3] A. Einstein and W. de Sitter, *On the relation between the expansion and the mean density of the universe*, Proc. Nat. Acad. Sci. **18** (1932) 213.

[4] A. Friedman, *On the curvature of space*, Z. Phys. **10** (1922) 377 [Gen. Rel. Grav. **31** (1999) 1991].

[5] A. Einstein, *Zum kosmologischen problem der allgemeinen relativitätstheorie*, Sitzungsb. König. Preuss. Akad. (1931) 142-152, reprinted and translated in *The Principle of Relativity* (Dover, 1952) p 175-188.

[6] C. O'Raifeartaigh and B. McCann, *Einstein's cosmic model of 1931 revisited: an analysis and translation of a forgotten model of the universe*, Eur. Phys. J. H **39** (2014) 63 [arXiv:1312.2192 [physics.hist-ph]].

[7] A. Sandage, L. M. Lubin and D. A. VandenBerg, *The age of the oldest stars in the local galactic disk from Hipparcos parallaxes of G and K subgiants*, Publ. Astron. Soc. Pac. **115** (2003) 1187 [astro-ph/0307128].

[8] H. E. Bond, E. P. Nelan, D. A. VandenBerg, G. H. Schaefer and D. Harmer, *HD 140283: A star in the solar neighborhood that formed shortly after the Big Bang*, Astrophys. J. **765** (2013) L12 [arXiv:1302.3180 [astro-ph.SR]].

[9] K.C. Jacobs, *Friedmann cosmological model with both radiation and matter*, Nature **215** (1967) 1156.

CHAPTER 8

The cosmological constant

I n this chapter we discuss a non-standard ingredient, not observed in ground laboratory experimental results, called cosmological constant. This was originally introduced by Einstein in 1917 with the motivation to obtain a static model where the standard gravitational attraction is balanced by a gravitational repulsion due to the presence of the (positive) cosmological constant. Though a static model was soon after ruled out by the discovery of the cosmological expansion made by Hubble in 1929, the role played by a cosmological constant within dynamical (Friedmann) models was intensively studied by Lemaitre and it was realised that it can provide an interesting solution to the age problem, as we will discuss in the next chapter.

8.1 FRIEDMANN EQUATION WITH A COSMOLOGICAL CONSTANT

In 1917 Einstein proposed the first cosmological model based on his new general theory of relativity, presented in its final version just the year before, marking the birth of modern cosmology [1]. He started from the idea that the model should respect the perfect cosmological principle, with no privileged time in the history of the universe and in particular without an origin of time. After all, at that time, a static universe appeared as the most attractive (and minimal) solution, since there was no experimental evidence supporting the expansion of the universe, not yet.

He had to face the same problem that Newton had almost three centuries earlier. From (his own!) Einstein's equations, he realised that as in the Newtonian case, gravity could only be attractive and matter could, therefore, never be distributed in a stable static configuration. On the other hand, he also noticed that the principle of equivalence and the request of covariance of the equations, were not forbidding an additional term in Einstein's equations (4.118). These could then be written more generally as

$$G^\mu_{\ \nu} \equiv R^\mu_{\ \nu} - \frac{1}{2}\delta^\mu_{\ \nu} \mathcal{R} - \delta^\mu_{\ \nu} \frac{\Lambda}{c^2} = \frac{8\pi G}{c^4} T^\mu_{\ \nu}, \tag{8.1}$$

where Λ is a constant, both in space and in time, called *cosmological constant*. The introduction of this term can potentially spoil the success of general relativity in reproducing the Newtonian description of gravity in the limit of weak and stationary field and for non-relativistic motions. As we have seen in Chapter 4, in this Newtonian limit, Einstein's equations reduce to the usual Poisson's equation Eq. (4.116) that would now get modified as

$$\nabla^2 \phi = 4\pi G \rho(\vec{r}) - \Lambda \,. \tag{8.2}$$

In particular, for $\Lambda > 0$, the new term on the right-hand side would now introduce a repulsive gravity contribution able, potentially, to balance the standard attractive gravity term. Clearly this term has to be sufficiently small not to produce deviations from Newtonian gravity so as to spoil its successful description of planetary and stellar dynamics.

The cosmological constant term can also be interpreted, instead of an additional term modifying Einstein's equations, as a contribution to the energy-momentum tensor from a non-standard fluid [2, 3]. If we indeed move the cosmological constant term on the the right-hand side of Einstein's equations (8.1), these can be recast as

$$G^\mu{}_\nu = \frac{8\pi G}{c^4}\left[(T_{\Lambda=0})^\mu{}_\nu + T^\mu_{\Lambda\,\nu}\right], \tag{8.3}$$

where we defined

$$T^\mu_{\Lambda\,\nu} \equiv \begin{pmatrix} \varepsilon_\Lambda & 0 & 0 & 0 \\ 0 & -p_\Lambda & 0 & 0 \\ 0 & 0 & -p_\Lambda & 0 \\ 0 & 0 & 0 & -p_\Lambda \end{pmatrix}, \quad \varepsilon_\Lambda \equiv \frac{\Lambda c^2}{8\pi G} = -p_\Lambda \tag{8.4}$$

and indicated with $(T_{\Lambda=0})^\mu{}_\nu$ the contribution to the energy-momentum tensor from standard fluids, such as matter and radiation, in the case of a vanishing cosmological constant.

Remembering that for an ideal fluid the energy-momentum tensor has to be of the form given by Eq. (4.73), the quantities ε_Λ and p_Λ can be identified with the energy density and pressure of a fluid described by the equation of state $p_\Lambda = w_\Lambda \varepsilon_\Lambda$, with $w_\Lambda = -1$. Gravitationally, a *cosmological constant term is completely equivalent and indistinguishable from a (non-standard) fluid with $p_\Lambda = -\varepsilon_\Lambda$* and for this reason we will refer to it as *cosmological constant-like fluid*. It is very interesting that within *quantum field theories*, such as the standard model, *vacuum energy density* provides a very well-motivated physical realisation of cosmological constant-like fluid.[1]

With such an interpretation of the cosmological constant, it is straightforward to write an extended version of the Friedmann equation (6.30) in the form

$$H^2 \equiv \left(\frac{\dot{a}}{a}\right)^2 = \frac{8\pi G}{3c^2}(\varepsilon_{\Lambda=0} + \varepsilon_\Lambda) - \frac{k c^2}{a^2 R_0^2}, \tag{8.5}$$

---

[1]However, unfortunately, an actual calculation of the value of the cosmological constant in quantum field theories represents one of the most formidable scientific challenges, since quantum field theories have predicted values that are many orders of magnitude bigger than the measured one, up to 122 orders of magnitude. Today we do not have a better solution than assuming that different contributions cancel out with incredible precision, translating into the greatest fine-tuning problem in the history of phsyics!

or, equivalently, as

$$H^2 \equiv \left(\frac{\dot{a}}{a}\right)^2 = \frac{8\,\pi\,G}{3\,c^2}\,\varepsilon_{\Lambda=0} - \frac{k\,c^2}{a^2\,R_0^2} + \frac{\Lambda}{3}\,, \tag{8.6}$$

where we are indicating with $\varepsilon_{\Lambda=0}$ the usual energy density one would have for $\Lambda = 0$. Correspondingly, the Friedmann equation written in terms of Ω_0 and H_0 (see Eq. (6.50)) becomes

$$\dot{a}^2 = H_0^2\,\Omega_0\,\frac{\varepsilon_{\Lambda=0} + \varepsilon_\Lambda}{\varepsilon_0}\,a^2 + H_0^2\,(1 - \Omega_0)\,, \tag{8.7}$$

where now $\Omega_0 = \Omega_{\Lambda=0,0} + \Omega_{\Lambda,0}$. It should be noticed from Eq. (8.4) that $\varepsilon_\Lambda = $ const. This result is completely consistent with the fluid equation for a cosmological constant-like fluid that reads like

$$\dot{\varepsilon}_\Lambda = -3\,\frac{\dot{a}}{a}\,(\varepsilon_\Lambda + p_\Lambda)\,, \tag{8.8}$$

from which, since $p_\Lambda = -\varepsilon_\Lambda$, it follows immediately $\dot{\varepsilon}_\Lambda = 0$ and therefore $\varepsilon_\Lambda = $ const. The acceleration equation Eq. (6.40) becomes now

$$\ddot{a} = -\frac{4\,\pi\,G}{3\,c^2}\,[(\varepsilon + 3p)_{\Lambda=0} - 2\,\varepsilon_\Lambda]\,a\,, \tag{8.9}$$

or equivalently

$$\ddot{a} = -\frac{4\,\pi\,G}{3\,c^2}\,(\varepsilon + 3p)_{\Lambda=0}\,a + \frac{\Lambda}{3}\,a\,. \tag{8.10}$$

8.2 EINSTEIN'S STATIC MODEL

The acceleration equation confirms that a positive value of Λ gives rise to a repulsive term, contrasting the standard attractive term. Therefore, one can conceive a static model where the two terms perfectly cancel each other out, the starting assumption of *Einstein's static model* (or Einstein universe) [1], from which its features can be derived.

First of all, let us assume that the standard fluid is just made of matter, so that $p_{\Lambda=0} = p_M = 0$ and $\varepsilon_{\Lambda=0} = \varepsilon_M$. As a first condition for a static universe one has to impose $\ddot{a} = 0$, and in this way one immediately finds the condition

$$\varepsilon_\Lambda = \frac{\varepsilon_M}{2} \quad \Leftrightarrow \quad \Lambda = \frac{4\,\pi\,G}{c^2}\,\varepsilon_M\,. \tag{8.11}$$

As a second condition, one has to impose that the expansion rate (the Hubble parameter) vanishes as well, so that the scale factor $a = a_E = $ const. Therefore, from the Friedmann equation,[2] one can write

$$\frac{8\,\pi\,G}{3\,c^2}\,(\varepsilon_M + \varepsilon_\Lambda) - \frac{k\,c^2}{a_E^2\,R_0^2} = 0\,. \tag{8.12}$$

---

[2]In this case the Friedmann equation cannot be written in the form (8.7) since $H = 0$ and one cannot introduce the critical energy density and the energy density parameters.

From this relation, using Eq. (8.11), one finds first of all that the curvature parameter $k = +1$ (as expected, since it has to be a closed universe) and then

$$R_E = \frac{c^2}{\sqrt{4\pi\,G\,\varepsilon_M}} = \frac{c}{\sqrt{\Lambda}}\,, \tag{8.13}$$

where $R_E \equiv a_E\,R_0$ is the radius of curvature in the Einstein model.

At first sight, Einstein's static solution seems to be an attractive option but, as pointed out by Eddington, it is highly unstable since a non-vanishing tiny deviation $\delta R = R - R_E \neq 0$ would immediately result either into an expansion or into a collapse. We will be back on this point in a more general way when we meet again Einstein's static model within a procedure able to provide a general classification of models with a cosmological constant.

8.3   THE DE SITTER MODEL

There is another model where the perfect cosmological principle (no privileged time) is realised, even though in a dynamical way. It is given by a flat universe containing only a cosmological constant-like fluid. In this case we can specialise the Friedmann equation (8.7) using $p_{\Lambda=0} = \varepsilon_{\Lambda=0} = k = 0$ and $\varepsilon = \varepsilon_\Lambda = \varepsilon_0 = \mathrm{const}$, obtaining

$$\dot{a} = H_0\,a\,, \tag{8.14}$$

where we considered only the expansion solution with positive sign. This model of course realises the special case of a one-fluid model with $p = -\varepsilon$ and $w = -1$, a cosmological constant-like fluid. In the previous chapter, when we studied one-fluid models with $p = w\,\varepsilon$, we said that this case deserved special consideration. Now we have also found a physical justification for such kind of model. The solution to Eq. (8.14) can be written as

$$a(t) = e^{H_0\,(t-t_0)}\,. \tag{8.15}$$

Clearly now the expansion is exponentially accelerated instead of decelerated as in the case of models dominated by standard fluids.

Even though this model is dynamical, there is still no privileged time and, therefore, the de Sitter model is a perfectly time homogeneous model with a constant expansion rate. This is possible since the expansion is self-similar, it never changes with time. Moreover there is no singularity. The model is, therefore, quite attractive and was proposed by de Sitter soon after Einstein's static model [4]. However, it cannot describe the current cosmological observations, in particular it was immediately realised that it could not reproduce correctly the cosmological redshifts. On the other hand, as we will see, the de Sitter model will be recovered within the current standard cosmological model describing all observations, as an asymptotic limit of the universe expansion both in its very early stage and in its final stage. We will indeed arrive at the conclusion that not only do current observations support the occurrence of an initial inflationary stage, approximating quite closely a de Sitter expansion regime prior to the onset of a radiation-dominated regime, but they also indicate that our universe has just recently entered a new de Sitter expansion stage.

8.4 LEMAITRE MODELS

A multi-fluid model containing a Λ-like component with $w_\Lambda = -1$ opens many new possibilities for cosmological scenarios compared to the matter-radiation case. Let us in particular consider models containing an admixture of matter, radiation and a Λ-like fluid. They were extensively studied by Lemaitre and for this reason they are called Lemaitre models [5].

As we discussed already, for a mixture of fluids, without exchange of energy among them, the solution for the total energy density coming from the fluid equation is simply given by the sum of the solutions for the individual components. Therefore, in our case this can be written as

$$\varepsilon(a) = \frac{\varepsilon_{R,0}}{a^4} + \frac{\varepsilon_{M,0}}{a^3} + \varepsilon_{\Lambda,0} \,, \tag{8.16}$$

where notice that the contribution from Λ is constant. The present value of the total energy density parameter can then be written, correspondingly, as the sum of three contributions,

$$\Omega_0 = \Omega_{R,0} + \Omega_{M,0} + \Omega_{\Lambda,0} \,, \tag{8.17}$$

where, as for matter and radiation, we defined $\Omega_{\Lambda,0} \equiv \varepsilon_{\Lambda,0}/\varepsilon_{c,0}$. The general Friedmann equation in a dimensionless form, Eq. (6.51), for this three-fluid model specialises into

$$\frac{\dot{a}^2(t)}{H_0^2} = a^2(t) \left(\frac{\Omega_{R,0}}{a^4} + \frac{\Omega_{M,0}}{a^3} + \Omega_{\Lambda,0} \right) + (1 - \Omega_0) \,. \tag{8.18}$$

We can now again apply the graphical method applied to the matter-radiation model discussed in the previous chapter. The general definition for the effective potential, Eq. (6.54), specialises into

$$V(a) = V_{RM\Lambda}(a) \equiv -a^2 \left(\frac{\Omega_{R,0}}{a^4} + \frac{\Omega_{M,0}}{a^3} + \Omega_{\Lambda,0} \right) \,. \tag{8.19}$$

Eq. (8.18) can then be written, specialising Eq. (6.55), as

$$E_{RM\Lambda}(t) \equiv \frac{\dot{a}^2(t)}{H_0^2} + V_{RM\Lambda}(a) = E_0 \,, \tag{8.20}$$

showing that $E_{RM\Lambda}(a)$ is an integral of motion, i.e., it remains constant during the dynamical evolution. This simple observation allows to classify in a systematic way all Lemaitre models.

As we already discussed in the case of the matter-radiation model, the expansion of the universe corresponds to considering trajectories for increasing a and with a fixed given value $E_{RM\Lambda}(a) = E_0$. The trajectories can never enter the unphysical region below the effective potential $V(a)$, since one would have $E_0 - V_{RM\Lambda}(a) = \dot{a}^2/H_0^2 < 0$.

Qualitatively the case of positive Λ is very different from the case of negative Λ

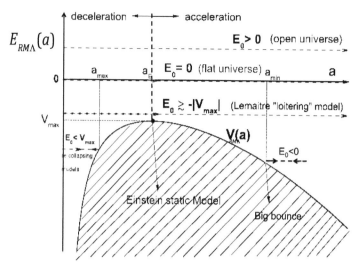

FIGURE 8.1: Classification of Lemaitre models, characterised by a mixture of radiation, matter and Λ based on the same graphical method shown in Fig. 7.3 for the matter-radiationuniverse.

and certainly phenomenologically much more interesting. For this reason here we focus only on this case, leaving the discussion of the case $\Lambda < 0$ to the reader (see Exercise 8.4).

The potential $V_{RM\Lambda}(a)$, with positive Λ, is plotted in Fig. 8.1. From this plot we can easily classify all possible cosmological models starting from some simple considerations. First of all notice that this time, contrarily to the matter-radiation case, the effective potential does not increase monotonically, but there is a particular value of the scale factor, corresponding to the same value a_E defined in the case of Einstein's static model, where it reaches a maximum value $V_{\max} \equiv V_{RM\Lambda}(a_E) < 0$. This feature implies a much richer variety of models compared to the matter-radiation case.

A second general consideration is that for $a < a_E$ one has a decelerated expansion regime, where the standard attractive gravitational contribution from matter and radiation dominates in the acceleration equation, while, for $a > a_E$, one has an accelerated expansion regime, where the repulsive gravitational term stemming from Λ dominates. For $a = a_E$ the acceleration vanishes, since for this value the attractive and repulsive terms exactly balance (as in the case of the static Einstein model discussed before).

It is then easy to classify the different possible models based on the values of E_0 and a_{in}.

- Let us start first from models with $E_0 > 0$ (open universe). Starting from an initial value of the scale factor $a_{\text{in}} < a_E$ there will be first a decelerated regime, for $a < a_E$. For $a = a_E$, at the peak of the potential, the expansion rate reaches its minimum and after that, for $a > a_E$ an accelerated regime will start and last indefinitely turning asymptotically into a de Sitter regime. Clearly if $a_{in} \geq a_E$ there is no decelerated regime.

- For a flat universe, $E_0 = 0$, the evolution is qualitatively the same as for the case of an open universe. We will specialise the discussion on this case in the next subsection, since it has a particular phenomenological importance.

- For a closed universe ($E_0 < 0$), there is now an important difference compared to the matter-radiation model. In the matter-radiation model for a closed universe the expansion stage is necessarily followed by a collapse. This time for $V_{max} < E_0 < 0$, corresponding to $1 < \Omega_0 < 1 + |V_{max}|$, one has a closed forever expanding universe with the same decelerated and accelerated stages as for the previous cases.

- Even for $E_0 < -|V_{max}|$ one can still have a forever expanding universe. Indeed suppose that the universe starts collapsing with an initial value of the scale factor $a_{in} > a_E$. There would then be a time when the universe *bounces* with a reaching a minimum value a_{min} and with the collapse phase turning into a subsequent forever expanding phase. This *Big Bounce* model has the clear advantage that there is no past singularity.[3] One can also again have, for $a_{in} < a_E$, re-collapsing models with a past singularity ($a = 0$) and a time when the scale factor reaches a maximum value a_{max} where the expansion stops and a collapsing phase follows.

- A model with $a_{in} \ll a_E$ and $E_0 = -|V_{max}| + |\delta E|$, i.e., with E_0 slightly above $-|V_{max}|$, is characterised by a long slow expansion phase about $a \simeq a_E$. This is the *loitering* (or *hesitating*) *model* [6] originally proposed by Lemaitre and later invoked in order to explain an observed star formation rate at redshifts $z \simeq 2$ much higher than in other epochs [7]. The loitering model for $z_E \equiv a_E^{-1} - 1 \simeq 2$ would indeed imply that the universe spent a much longer time at redshifts about z_E than in other intervals of redshift. A variation of this model, also proposed by Lemaitre, assumed in addition a tiny value of a_{in}, very close to the singularity. This was motivated by the possibility to explain the initial expansion rate as the effect of a sort of radiative decay of the initial universe to be regarded as a decaying *primeval atom* [8]. The model of the primeval atom marks the birth of *Big Bang models*.

- There is a very special model corresponding to a precise choice $E_0 = -|V_{max}|$. In this case, starting with $a_{in} < a_E$, the universe expansion would asymptotically stop for $a \to a_E$, with the universe sitting right on the top of the potential. This asymptotic limit would correspond to Einstein's static model. However, it should be clear now why this model is highly unstable: a tiny deviation from $a = a_E$ or from $E_0 = -|V_{max}|$ would result either into a collapse or into an expansion, as first noticed by Eddington. In 1927 Lemaitre proposed

---

[3]Of course one could also assume a straight initial expansion with $a_{in} > a_E$ without Big Bounce.

such a kind of expanding model with $a_{\text{in}} = a_E + |\delta a|$, with a very slow initial accelerated expansion becoming gradually faster and lasting forever [5]. In 1930 Eddington showed explicitly the instability of Einstein's model and how Lemaitre's proposal was the natural consequence [9]. This is today known as the *Lemaitre–Eddington model.*

An important question is whether among this broad class of Lemaitre models, there is one that can correctly reproduce the cosmological observations. We will answer this question in Chapter 10, finding that there is indeed a specific Lemaitre model supported by current cosmological observations and that this model is flat ($\Omega_0 = 1$ and $E_0 = 0$). For this reason it is useful to give more details on this particularly significant subclass of Lemaitre models.

8.4.1 Flat Lemaitre models

If we consider Eq. (8.16), it is evident that going sufficiently back in the past ($a \to 0$) one has a radiation-dominated regime, whereas for sufficiently high values of a a Λ-dominated regime, characterised by a de Sitter (exponential) expansion, is realised. Let us assume that between these two asymptotical regimes a matter-dominated regime takes place as well, in a way that one has a three-regime model.

During the initial radiation-dominated regime one has

$$a(t) \simeq a_{\text{eq}} \left(\frac{t}{t_{\text{eq}}} \right)^{\frac{1}{2}} , \quad (t < t_{\text{eq}}) . \tag{8.21}$$

The initial radiation-dominated regime ends at the matter-radiation equality time t_{eq}. It corresponds to a value of the scale factor given exactly by the same expression Eq. (7.43) found already when we studied the matter-radiation model.

At $t = t_{\text{eq}}$ the matter-dominated regime starts and therefore during this regime the scale factor behaviour is approximately given by

$$a(t) \simeq a_{\text{eq}} \left(\frac{t}{t_{\text{eq}}} \right)^{\frac{2}{3}} \quad \left(t_{\text{eq}} < t < t_{\text{eq}}^{M\Lambda} \right) . \tag{8.22}$$

The matter-dominated regime ends at $t_{\text{eq}}^{M\Lambda}$, the matter-Λ equality time, when the matter contribution and the Λ contribution to the total energy density equal each other, explicitly

$$\frac{\varepsilon_{M,0}}{(a_{\text{eq}}^{M\Lambda})^3} = \varepsilon_{\Lambda,0} , \tag{8.23}$$

where $a_{\text{eq}}^{M\Lambda} \equiv a(t_{\text{eq}}^{M\Lambda})$. One easily finds

$$a_{\text{eq}}^{M\Lambda} = \left(\frac{\Omega_{M,0}}{\Omega_{\Lambda,0}} \right)^{\frac{1}{3}} . \tag{8.24}$$

The corresponding redshift at the matter-Λ equality time is then given by

$$z_{\text{eq}}^{M\Lambda} = \frac{1}{a_{\text{eq}}^{M\Lambda}} - 1 = \left(\frac{\Omega_{\Lambda,0}}{\Omega_{M,0}} \right)^{\frac{1}{3}} - 1 . \tag{8.25}$$

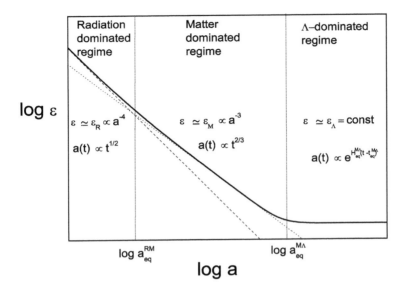

FIGURE 8.2: Flat Lemaitre model. The solid line is the dependence of the energy density ε on the scale factor a (notice the log-log plot). The dashed line is the asymptotic behaviour in the radiation-dominated regime given by $\varepsilon \sim \varepsilon_R \propto a^{-4}$, the dotted line is the behaviour in the intermediate matter-dominated regime, $\varepsilon \sim \varepsilon_M \propto a^{-3}$. In the Λ dominated regime finally $\varepsilon \sim \varepsilon_\Lambda = $ const.

Finally, a Λ-dominated regime starts and the universe expansion turns gradually into a de Sitter exponential expansion described by

$$a(t) \simeq a_{\rm eq}^{M\Lambda} \, e^{H_{\rm eq}^{M\Lambda} \, (t-t_{\rm eq}^{M\Lambda})} \qquad (t > t_{\rm eq}^{M\Lambda}), \qquad (8.26)$$

where $H_{\rm eq}^{M\Lambda}$ is the Hubble parameter at $t_{\rm eq}^{M\Lambda}$. Within this approximation of instantaneous transition from one regime to another, one has $H_{\rm eq}^{M\Lambda} = H_0$. A more precise calculation is able to specify more precisely the variation of the expansion rate from the matter-Λ equality time to the present.

We can finally see that the initial assumption of the existence of a matter-dominated regime implies the condition

$$a_{\rm eq} < a_{\rm eq}^{M\Lambda} \quad \Leftrightarrow \quad \Omega_{M,0} \gtrsim \Omega_{R,0}^{\frac{3}{4}} \, \Omega_{\Lambda,0}^{\frac{1}{4}} . \qquad (8.27)$$

If this condition is not verified, then there would not be a matter-dominated regime but simply the radiation-dominated regime would be followed directly by the Λ-dominated regime. These results are summarised in Fig. 8.2. We will see in the next chapter that Lemaitre models are able to provide a solution to the age problem that we encountered discussing flat models without a cosmological constant.

EXERCISES

Exercise 8.1 *What is the dimensionality of the cosmological constant Λ?*

Exercise 8.2 *Find the value of the matter energy density ε_M (in MeV/m^3) for an Einstein Universe with radius $R_E = 1$ Gpc. Find the corresponding value of the cosmological constant Λ (in yr^{-2}).*

Exercise 8.3 *Consider the expression*

$$V(a) = V_{RM\Lambda}(a) \equiv -a^2 \left(\frac{\Omega_{R,0}}{a^4} + \frac{\Omega_{M,0}}{a^3} + \Omega_{\Lambda,0} \right) \qquad (8.28)$$

for the effective potential in a Lemaitre model with an admixture of radiation, matter and a fluid with $p_\Lambda = -\varepsilon_\Lambda$ corresponding to a cosmological constant in the Einstein equations. Find an expression for $V_{\max}$ and for a_E.

Exercise 8.4 *What cosmological models would one obtain if instead of a positive cosmological constant term one would consider a negative one? Would models with a big bounce be possible in this case? Describe the situation graphically showing the dynamics in a plot where the scale factor is on the x axis and the effective energy on the y-axis.*

BIBLIOGRAPHY

[1] A. Einstein, *Cosmological considerations in the general theory of relativity*, Sitzungsber. Preuss. Akad. Wiss. Berlin (Math. Phys.) **1917** (1917) 142.

[2] G. Lemaitre, *Evolution of the expanding universe*, Proc. Nat. Acad. Sci. **20** (1934) 12.

[3] A. D. Sakharov, *Vacuum quantum fluctuations in curved space and the theory of gravitation*, Sov. Phys. Dokl. **12** (1968) 1040 [Dokl. Akad. Nauk Ser. Fiz. **177** (1967) 70] [Sov. Phys. Usp. **34** (1991) 394] [Gen. Rel. Grav. **32** (2000) 365].

[4] W. de Sitter, *Einstein's theory of gravitation and its astronomical consequences, Third Paper*, Mon. Not. Roy. Astron. Soc. **78** (1917) 3.

[5] G. Lemaitre, *A homogeneous universe of constant mass and increasing radius accounting for the radial velocity of extra-galactic Nebulae*, Monthly Notices of the Royal Astronomical Society, Vol. 91, p.483-490 (1931), translated from Lemaitre, Georges (1927), *Un univers homogene de masse constante et de rayon croissant rendant compte de la vitesse radiale des nebuleuses extra-galactiques*, Annales de la Societe Scientifique de Bruxelles A47: 49-56

[6] G. Lemaitre, *The expanding universe*, Mon. Not. Roy. Astron. Soc. **91** (1931) 490.

[7] V. Petrosian, E. Salpeter and P. Szekeres, *Quasi-stellar objects in universes with non-zero cosmological constant*, Astrophys. J. **147** (1967) 1222.

[8] G. Lemaitre, *The beginning of the world from the point of view of quantum theory*, Nature 127, 706-706 (09 May 1931).

[9] A.S. Eddington, *On the instability of Einstein's spherical world*, Monthly Notices of the Royal Astronomical Society, **90** (1930) 668.

Age of the universe

I n this chapter we discuss a general procedure to solve the Friedmann equation, finding a general integral expression for the age of the universe, depending on the observable cosmological parameters, in particular H_0 and Ω_0. In this way we manage to identify two different solutions to the age problem: the first one is an open universe without cosmological constant (i.e., a purely matter model), the second one is a flat Lemaitre model with an admixture of matter and Λ-like fluid.[1]

9.1 A GENERAL EXPRESSION FOR THE AGE OF THE UNIVERSE

We found simple analytic expressions for the age of the universe t_0 and for the scale factor time dependence $a(t)$ in a few simple cases:

(i) empty universe: $a(t) = t/t_0$ with $t_0 = H_0^{-1}$;

(ii) flat matter universe: $a(t) = (t/t_0)^{\frac{2}{3}}$ with $t_0 = 2\,H_0^{-1}/3$;

(iii) flat radiation universe: $a(t) = (t/t_0)^{\frac{1}{2}}$ with $t_0 = H_0^{-1}/2$;

(iv) flat models with $p = w\,\varepsilon$ and $w = \text{const} > -1$:

$$a(t) = \left(\frac{t}{t_0}\right)^{\frac{2}{3\,(1+w)}} \quad \text{with} \quad t_0 = \frac{2\,H_0^{-1}}{3\,(1+w)}. \tag{9.1}$$

We noticed that cases (ii) and (iii), corresponding to standard fluids, suffer from an age problem since, for $H_0 \simeq 70\,\text{km}\,\text{s}^{-1}\,\text{Mpc}^{-1}$, one finds $t_0 \lesssim 10\,\text{Gyr}$, in disagreement with the estimated age of the oldest stars in globular clusters that place a very robust lower bound $t_0 \gtrsim 12\,\text{Gyr}$ [2, 3, 4].

The empty universe seems to provide a possible solution to the age problem but one could object that it is unrealistic, since we do observe matter and radiation in

---

[1]These were the two most popular solutions in the mid 1990s before the discovery of the acceleration of the universe and of CMB acoustic peaks that clearly support the second one. A nice review on the uncertain situation before the advent of these important results can be found in [1].

the universe. However, the empty model can be actually regarded as the asymptotic limit of open models for $\Omega_0 \rightarrow 0$, independently of the nature of the fluid and in this sense it suggests a perfectly reasonable solution.

In case (iv), we also noticed that for a fluid with $w \simeq -1/3$ one would also find, as in the case of the empty model, $t_0 \simeq H_0^{-1} \simeq 14\,\text{Gyr}$, in agreement with the age of the oldest stars. On the other hand, it is not easy to justify physically the existence of a fluid with such an equation of state. We want now to face the age problem in a systematic way, finding a general expression for the age of the universe.

Let us start from Eq. (6.55), the Friedmann equation in a dimensionless form expressed in terms of the cosmological parameters at present, the effective potential $V(a)$ and the total effective energy E_0. As discussed, this form explicitly shows how the Friedmann equation can be mathematically regarded as a one-dimensional conservative system, like for a particle moving in a one-dimensional potential.[2] We can then employ a well-known procedure to solve this kind of differential equation.

Taking the square root of both sides of the Friedmann equation, we easily obtain

$$\frac{da}{dt} = \pm H_0 \sqrt{E_0 - V(a)}, \tag{9.2}$$

where the plus sign holds for expansion stages and the minus sign for collapse stages. From this equation we can then express the time differential as

$$dt = \pm \frac{H_0^{-1}\, da}{\sqrt{E_0 - V(a)}}, \tag{9.3}$$

and upon integration one finds

$$t = t_{\text{in}} \pm H_0^{-1} \int_{a_{\text{in}}}^{a} \frac{da'}{\sqrt{E_0 - V(a')}}, \tag{9.4}$$

where t_{in} is some initial time and we defined $a_{\text{in}} \equiv a(t_{\text{in}})$. This is a very general solution that gives $t = t(a)$ during a period of the universe dynamics with a fixed sign of the velocity. If there is a change of sign, like for example in a closed universe with an initial expansion followed by a collapse or, conversely, in models with a big bounce, then one has to sum over two different periods with a different sign. Notice moreover that after having obtained a relation $t = t(a)$, this can be inverted yielding a relation $a = a(t)$. This procedure is dictated by the fact that the fluid equation provides a relation $\varepsilon = \varepsilon(a)$ and not a relation $\varepsilon = \varepsilon(t)$, so that one has first to find $t(a)$ and then, inverting, $a(t)$.

Let us now specialise our discussion to *models with an initial singularity* $(a(t_s) = 0)$ and such that the universe monotonically expands from the initial time to the present. As already pointed out, we conventionally set the origin of time at the singularity, so that $t_s = 0$ and $a(t_s = 0)$, and define *cosmic time* t as the time

---

[2]This is in agreement with the fact, discussed in Chapter 3, that in the special case $p = 0$ the Friedmann equation can be derived within a Newtonian model, though we noticed that this does not provide a satisfactory physical picture for different reasons.

elapsed from the singularity $(t_{\text{in}} = 0)$.[3] In this way for models with a singularity one finds for the cosmic time

$$t = H_0^{-1} \int_0^a \frac{da'}{\sqrt{E_0 - V(a')}} . \tag{9.5}$$

and for the *age of the universe*

$$t_0 = H_0^{-1} f_\varepsilon(\Omega_0), \quad \text{with} \quad f_\varepsilon(\Omega_0) \equiv \int_0^1 \frac{da}{\sqrt{E_0 - V(a)}} . \tag{9.6}$$

One can easily check that this general integral expression for t_0 reproduces correctly the results for the four simple cases reviewed at the beginning. For example in case (iv) one has $E_0 = 0$ and plugging the expression Eq. (7.32) for $\varepsilon(a)$ into the definition of the effective potential Eq. (6.54), one finds $V(a) = a^{-(1+3\,w)}$ and from this

$$t_0 = H_0^{-1} \int_0^1 da\, a^{\frac{(1+3\,w)}{2}} = \frac{2}{3\,(1+w)} H_0^{-1}, \tag{9.7}$$

confirming the result in Eq. (7.36).

9.2   LEMAITRE MODELS

Let us now consider the most interesting case of Lemaitre models containing an admixture of matter, radiation and a cosmological constant. For these fluids, as we have seen, we can write

$$V(a) = V_{RM\Lambda}(a) \equiv -\left(\frac{\Omega_{R,0}}{a^2} + \frac{\Omega_{M,0}}{a} + \Omega_{\Lambda,0}\, a^2 \right) . \tag{9.8}$$

In this way Eq. (9.5) for the cosmic time becomes explicitly

$$t(a) = H_0^{-1} \int_0^a \frac{da'}{\sqrt{1 - \Omega_0 + \frac{\Omega_{R,0}}{a'^2} + \frac{\Omega_{M,0}}{a'} + \Omega_{\Lambda,0}\, a'^2}}, \tag{9.9}$$

while the expression for the age of the universe Eq. (9.6) specialises into

$$t_0 = H_0^{-1} f_{RM\Lambda}(\Omega_{R,0}, \Omega_{M,0}, \Omega_{\Lambda,0}), \tag{9.10}$$

where we defined

$$f_{RM\Lambda}(\Omega_{R,0}, \Omega_{M,0}, \Omega_{\Lambda,0}) \equiv \int_0^1 \frac{da}{\sqrt{1 - \Omega_0 + \frac{\Omega_{R,0}}{a^2} + \frac{\Omega_{M,0}}{a} + \Omega_{\Lambda,0}\, a^2}} . \tag{9.11}$$

It is simple to verify that this equation reproduces the special results that we obtained already:

---

[3]Let us remember that the de Sitter model is excluded by this class of models since it has no singularity.

- empty universe ($\Omega_0 = \Omega_{R,0} = \Omega_{M,0} = \Omega_{\Lambda,0} = 0$) $\Rightarrow t_0 H_0 = 1$;

- flat matter universe ($\Omega_0 = \Omega_{M,0} = 1 \Rightarrow \Omega_{\Lambda,0} = \Omega_{R,0} = 0$) $\Rightarrow t_0 H_0 = 2/3$;

- flat radiation universe ($\Omega_0 = \Omega_{R,0} = 1 \Rightarrow \Omega_{M,0} = \Omega_{\Lambda,0} = 0$) $\Rightarrow t_0 H_0 = 1/2$.

Let us assume that the condition Eq. (8.27) holds and that a matter-dominated regime indeed occurs between the radiation and Λ-dominated regime. The age of the universe can then be decomposed into two contributions corresponding mathematically to two contributions to the integral $f_{RM\Lambda}(\Omega_{R,0}, \Omega_{M,0}, \Omega_{\Lambda,0})$, namely

$$f_{RM\Lambda}(\Omega_{R,0}, \Omega_{M,0}, \Omega_{\Lambda,0}) = \int_0^{a_{eq}} \frac{da}{\sqrt{1 - \Omega_0 + \frac{\Omega_{R,0}}{a^2} + \frac{\Omega_{M,0}}{a} + \Omega_{\Lambda,0} a^2}}$$

$$+ \int_{a_{eq}}^1 \frac{da}{\sqrt{1 - \Omega_0 + \frac{\Omega_{R,0}}{a^2} + \frac{\Omega_{M,0}}{a} + \Omega_{\Lambda,0} a^2}}.$$

The first contribution gives the duration of the radiation-dominated regime t_{eq}, while the second contribution is the remaining interval of time, $t_0 - t_{eq}$, elapsed from the matter-radiation equality time until the present time, explicitly. We can then approximately calculate these two contributions neglecting the matter energy density in the first term and the radiation energy density in the second term, explicitly

$$t_{eq} \simeq H_0^{-1} \int_0^{a_{eq}} \frac{da}{\sqrt{1 - \Omega_0 + \frac{\Omega_{R,0}}{a^2}}},$$

$$t_0 - t_{eq} \simeq H_0^{-1} \int_{a_{eq}}^1 \frac{da}{\sqrt{1 - \Omega_0 + \frac{\Omega_{M,0}}{a} + \Omega_{\Lambda,0} a^2}}. \tag{9.12}$$

Since, as we will discuss in the next chapter, it is found experimentally $\Omega_{R,0} \lll \Omega_{M,0}$ implying $a_{eq} \lll 1$, one has $t_{eq} \lll t_0$. In this way the age of the universe can always be calculated, with good approximation, neglecting the duration of the initial radiation-dominated regime, i.e.,

$$t_0 \simeq H_0^{-1} f_{M\Lambda}(\Omega_{M,0}, \Omega_{\Lambda,0}), \quad f_{M\Lambda}(\Omega_{M,0}, \Omega_{\Lambda,0}) \equiv \int_0^1 \frac{da}{\sqrt{1 - \Omega_0 + \frac{\Omega_{M,0}}{a} + \Omega_{\Lambda,0} a^2}},$$

$$\tag{9.13}$$

where notice that, by definition, $f_{M\Lambda}(\Omega_{M,0}, \Omega_{\Lambda,0}) = f_{RM\Lambda}(0, \Omega_{M,0}, \Omega_{\Lambda,0})$.

9.2.1 Matter universe

Let us first consider models without a cosmological constant but with a generic value of Ω_0 (matter universe) so that $\Omega_0 \simeq \Omega_{M,0}$ (we are neglecting the radiation

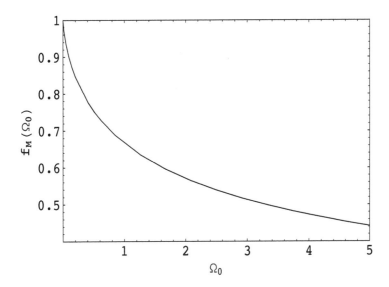

FIGURE 9.1: Plot of the function $f_M(\Omega_0)$ defined in Eq. (9.14).

contribution). Eq. (9.10) for the age of the universe specialises then into

$$t_0\,H_0 \simeq f_M(\Omega_0) \equiv \int_0^1 \frac{da}{\sqrt{1 - \Omega_0 + \frac{\Omega_0}{a}}}\,. \tag{9.14}$$

It is possible to derive an explicit analytical expression of this integral (see Exercise 9.1), finding for an open universe

$$f_M(\Omega_0 < 1) = \frac{1}{1 - \Omega_0} - \frac{\Omega_0\,\mathrm{arcsinh}\left[\sqrt{\frac{1-\Omega_0}{\Omega_0}}\right]}{(1 - \Omega_0)^{3/2}} \tag{9.15}$$

and for a closed universe

$$f_M(\Omega_0 > 1) = \frac{\Omega_0\,\mathrm{arcsin}\left[\sqrt{\frac{\Omega_0-1}{\Omega_0}}\right]}{(\Omega_0 - 1)^{3/2}} - \frac{1}{\Omega_0 - 1}\,. \tag{9.16}$$

A plot of the function $f_M(\Omega_0)$ is shown in the figure. One can see that for $\Omega_0 \to 0$ the empty universe limit $f_M(0) = 1$ is recovered, while for $\Omega_0 = 1$ the value for the flat case (the Einstein–de Sitter model), $f_M(1) = 2/3$, is recovered. This shows that the age problem can be solved considering an open universe and that a reasonable age of the universe compatible with the estimated age of oldest stars ($t_0 \gtrsim 12\,\mathrm{GeV}$) is obtained for $\Omega_0 \lesssim 0.2$. However, as we will see in the next chapter, the current cosmological observations strongly support a flat universe and, therefore, this solution to the age problem is currently ruled out.

9.2.2 Flat Lemaitre model

Let us then look for a solution to the age problem within a flat universe. In this case we have to solve the integral

$$f_{M\Lambda}(\Omega_{M,0}, \Omega_{\Lambda,0}) \equiv \int_0^1 \frac{da}{\sqrt{\frac{\Omega_{M,0}}{a} + \Omega_{\Lambda,0} \, a^2}}, \tag{9.17}$$

with $\Omega_{M,0} + \Omega_{\Lambda,0} = 1$. In this way there is only one independent parameter that can be chosen, for example, to be $\Omega_{\Lambda,0}$.

This integral can be calculated analytically in the following way. First of all one can notice that, remembering the expression Eq. (8.24) for the scale factor at the matter-Λ equality time $a_{eq}^{M\Lambda}$, it can be recast as

$$f_{M\Lambda}(1 - \Omega_{\Lambda,0}, \Omega_{\Lambda,0}) \equiv f(\Omega_{\Lambda,0}) \equiv \frac{1}{\sqrt{\Omega_{\Lambda,0}}} \int_0^{1/a_{eq}^{M\Lambda}} \frac{dx}{x\sqrt{1 + x^{-3}}}, \tag{9.18}$$

where we defined $x \equiv a/a_{eq}^{M\Lambda}$. One can then perform the change of variable $x = y^{2/3}$, so that the integral becomes

$$f(\Omega_{\Lambda,0}) = \frac{2}{3\sqrt{\Omega_{\Lambda,0}}} \int_0^{(a_{eq}^{M\Lambda})^{-3/2}} \frac{dy}{\sqrt{1 + y^2}}. \tag{9.19}$$

Finally, with a third change of variable $y = \sinh \alpha$, one finds

$$f(\Omega_{\Lambda,0}) = \frac{2}{3\sqrt{\Omega_{\Lambda,0}}} \operatorname{arcsinh}[(a_{eq}^{M\Lambda})^{-3/2}]. \tag{9.20}$$

Using now $\operatorname{arcsinh} x = \ln(x + \sqrt{1 + x^2})$, and recalling the expression Eq. (8.24) for $a_{eq}^{M\Lambda}$, one finds eventually

$$f(\Omega_{\Lambda,0}) = \frac{2}{3\sqrt{\Omega_{\Lambda,0}}} \ln\left[\frac{1 + \sqrt{\Omega_{\Lambda,0}}}{\sqrt{1 - \Omega_{\Lambda,0}}}\right]. \tag{9.21}$$

This expression is plotted in Fig. 9.2. One can see that it correctly reproduces the results

$$f(\Omega_{\Lambda,0} = 0) = 2/3 \quad \text{and} \quad f(\Omega_{\Lambda,0} = 1) = \infty. \tag{9.22}$$

The second result can be understood considering that the condition $\Omega_{\Lambda,0} = 1$ implies $\Omega_{M,0} = 0$ and therefore one recovers the pure Λ-dominated model (the de Sitter model), where the age of the universe is basically infinite since there is no singularity.

The general result given by Eq. (9.21) is plotted in Fig. 9.2. It can be noticed that for $\Omega_{\Lambda,0} \simeq 0.7$ one obtains again[4] $t_0 \simeq H_0^{-1} \simeq 14 \, \text{Gyr}$, therefore solving the

---

[4]We have already encountered two models with $t_0 = H_0^{-1}$: the empty model and a model with $p/\varepsilon = -1/3$.

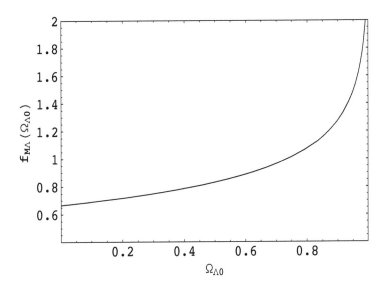

FIGURE 9.2: Plot of the function $f_{M\Lambda}(\Omega_{\Lambda,0})$ defined in Eq. (9.18).

age problem. As we will see in the next chapter, this will prove to be the value that is supported by other experimental observations as well and what is nowadays accepted as quite a robust experimental result.

After having calculated the age of the universe, one can also calculate the matter-Λ equality time given by

$$t_{eq}^{M\Lambda} = H_0^{-1} \int_0^{a_{eq}^{M\Lambda}} \frac{da}{\sqrt{\frac{\Omega_{M,0}}{a} + \Omega_{\Lambda,0}\, a^2}}. \tag{9.23}$$

Here one could approximate the integral neglecting the contribution from Λ and writing

$$t_{eq}^{M\Lambda} \simeq H_0^{-1} \int_0^{a_{eq}^{M\Lambda}} da\, \sqrt{\frac{a}{\Omega_{M,0}}}. \tag{9.24}$$

With some easy passages one then arrives at the expression

$$t_{eq}^{M\Lambda} \simeq \frac{2\, H_0^{-1}}{3\, \sqrt{\Omega_{\Lambda,0}}}. \tag{9.25}$$

In a precise calculation the cosmological constant has to be taken into account in the denominator of Eq. (9.23). The calculation proceeds along the same line followed to calculate $f_{M\Lambda}(\Omega_{\Lambda,0})$ with the only difference being that now the upper extreme of integration is $a_{eq}^{M\Lambda}$ and instead of Eq. (9.25) one obtains more precisely

$$t_{eq}^{M\Lambda} = \frac{2\, H_0^{-1}}{3\, \sqrt{\Omega_{\Lambda,0}}} \operatorname{arcsinh}(1) = \frac{2\, H_0^{-1}}{3\, \sqrt{\Omega_{\Lambda,0}}} \ln(1 + \sqrt{2})\,, \tag{9.26}$$

about 10% less than the previous approximated result.

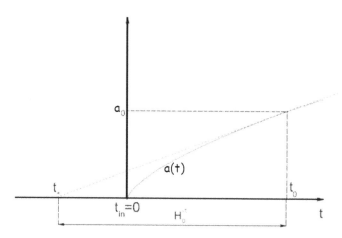

FIGURE 9.3: Geometrical relation between H_0^{-1} and t_0. The solid line is the relation $R(t)$, while the dashed line is the tangent to $R(t)$ in t_0.

9.3 SUMMARISING GRAPHICAL DESCRIPTION AND ΛCDM PREDICTION

Can we understand in a nutshell the previous results on the age of the universe and in particular the solutions to the age problem? There is a nice summarising graphical method that allows that. If one plots $a(t)$, one can notice that

$$a_0 = \dot{a}_0 \left(t_0 - t_\star \right), \tag{9.27}$$

where $t_\star$ is defined as that particular time, either positive or negative, where the tangent to $a(t)$ intercepts the time-axis (see Fig. 9.3). Therefore, $H_0^{-1} = (a_0/\dot{a}_0)$ is nothing else that the difference between t_0 and $t_\star$, explicitly

$$H_0^{-1} = t_0 - t_\star. \tag{9.28}$$

In particular, if $t_\star = 0$ then $t_0 = H_0^{-1}$ (or equivalently $f_\varepsilon = 1$). This special case corresponds to a uniformly expanding universe ($\dot{a} = $ const) and represents a borderline between ever-decelerating models with $f_\varepsilon < 1$ and ever-accelerating models with $f_\varepsilon > 1$. The case of a flat Lemaitre model[5] is then a hybrid model where in the first deceleration stage one has $f_\varepsilon < 1$ and in the acceleration stage, when Λ dominates, one has eventually $f_\varepsilon > 1$. There must be clearly a transient intermediate period where $f_\varepsilon \simeq 1$. In this model it is then quite a fine-tuned coincidence to have $t_0 \simeq H_0^{-1} \simeq 14\,\mathrm{Gyr}$. This happens when the cosmological constant has started to dominate the expansion only relatively recently (in cosmological terms!).

Curiously this is exactly what happens in the current ΛCDM model, currently supported by the observations, considering that its expansion history is described by a flat Lemaitre model after an initial inflationary stage. In Fig. 9.4 we have plotted in the top panel the cosmic time t as a function of redshift $z = a^{-1} - 1$ and in the bottom panel the evolution of the scale factor $a(t)$ for the best fit values of

---

[5]More generally, this occurs for Lemaitre models with $E_0 > -|V_{\mathrm{max}}|$ (see Fig. 8.1).

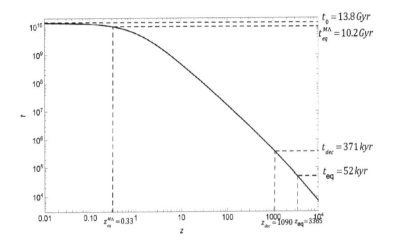

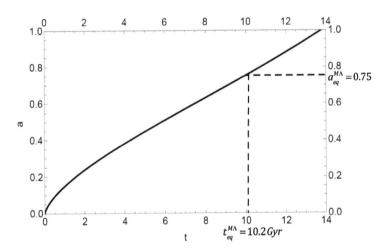

FIGURE 9.4: Cosmic time as a function of redshift (upper panel) and evolution of the scale factor with time (lower panel) calculated from Eq. (9.9) with $\Omega_0 = 1$, $\Omega_{M,0} = 0.308$, $\Omega_{\Lambda,0} = 0.692$, $\Omega_{R,0} = 0.915 \times 10^{-4}$, $H_0^{-1} = 14.4\,\mathrm{Gyr}$ (best CMB fit parameters from *Planck* collaboration [5] within the ΛCDM model). We have also indicated redshift and age of the universe at some special times in the history of the universe.

$\Omega_{R,0}, \Omega_{M,0}$ and $\Omega_{\Lambda,0}$ indicated in the figure caption and for $\Omega_0 = 1$. This choice turns out to give the right solution to the age problem as suggested by the experimental results. In particular it can be noticed from the plot of $a(t)$ how indeed with these values one has approximately $t_\star \simeq 0$ implying $t_0 \simeq H_0^{-1}$.

We have of course now to understand how the observations pin down this particular model. It is therefore time to turn back to the cosmological observations and see what kind of information we can get on the cosmological parameters.

EXERCISES

Exercise 9.1 *Solve the integral*

$$t_0 H_0 = f_M(\Omega_0) \equiv \int_0^1 \frac{da}{\sqrt{\frac{\Omega_0}{a} + (1 - \Omega_0)}} \tag{9.29}$$

for the age of the Universe of a matter Universe with arbitrary Ω_0, deriving the analytic expressions Eqs. (9.15) and (9.16) given in the text.

Exercise 9.2 *Consider an expanding, positively curved Universe containing only a cosmological constant ($\Omega_0 = \Omega_{\Lambda,0} > 1$). Show that such a Universe underwent a 'Big Bounce' at a value of the scale factor given by*

$$a_{\text{bounce}} = \left(\frac{\Omega_0 - 1}{\Omega_0} \right)^{1/2}. \tag{9.30}$$

Show that in this model the scale factor depends on time as

$$a(t) = a_{\text{bounce}} \cosh[\sqrt{\Omega_0}\, H_0\, (t - t_{\text{bounce}})], \tag{9.31}$$

where t_{bounce} is the time at which the Big Bounce occurred. What is the time elapsed since the Big Bounce, $t_0 - t_{\text{bounce}}$, expressed in terms of Ω_0 and H_0?

BIBLIOGRAPHY

[1] D. N. Schramm, FERMILAB-CONF-97-007-A.

[2] B. Chaboyer, P. J. Kernan, L. M. Krauss and P. Demarque, *A lower limit on the age of the universe*, Science **271** (1996) 957 [astro-ph/9509115].

[3] A. Sandage, L. M. Lubin and D. A. VandenBerg, *The age of the oldest stars in the local Galactic disk from Hipparcos parallaxes of G and K subgiants*, Publ. Astron. Soc. Pac. **115** (2003) 1187 [astro-ph/0307128].

[4] H. E. Bond, E. P. Nelan, D. A. VandenBerg, G. H. Schaefer and D. Harmer, *HD 140283: A star in the solar neighborhood that formed shortly after the Big Bang*, Astrophys. J. **765** (2013) L12 [arXiv:1302.3180 [astro-ph.SR]].

[5] P. A. R. Ade *et al.* [Planck Collaboration], *Planck 2015 results. XIII. Cosmological parameters*, arXiv:1502.01589 [astro-ph.CO].

CHAPTER 10

Expansion history of the universe

I n this chapter we show how the determination of the luminosity distance of astronomical sources at large redshifts has made it possible to trace the expansion of the universe back to billions of years ago and, consequently, to establish that this is accelerating at present. When this result is combined with the flatness of the universe from the observation of CMB anisotropies, the emerging cosmological model is a flat Lemaitre model with a mixture of radiation, matter and a positive cosmological constant. In this way the current acceleration of the expansion is interpreted as the consequence of the dominant contribution of the cosmological constant to the present total energy density. We start the chapter discussing an interesting question: how far can we observe the universe? An answer to this question requires the introduction of an important concept in cosmology: the horizon of the universe.

10.1 THE OBSERVABLE UNIVERSE

How big is the observable universe? Or in other words: how far away are the most distant points of the universe from which we can receive a signal?

We have seen that a light signal emitted by an astronomical source at time t_{em} obeys the usual condition for the propagation of light, $ds = 0$. Using the FRW metric we have then obtained that a signal traveling from the source to us obeys the relation Eq. (6.14) and from this, multiplying both sides by R_0, we obtained the relation Eq. (6.22).

The latter has a very important implication: in a universe with an origin at time t_{in}, corresponding to an initial value of the scale factor a_{in}, an observer at time $t > t_{\text{in}}$ cannot receive any light signal from sources located beyond the *horizon distance* defined as

$$d_H(t_{\text{in}}, t) \equiv a(t) \int_{t_{\text{in}}}^{t} \frac{c\, dt'}{a(t')}. \tag{10.1}$$

In particular, at the present time we cannot receive signals beyond the present horizon distance

$$d_{H,0}(t_{\text{in}}) \equiv \int_{t_{\text{in}}}^{t_0} \frac{c\,dt}{a(t)}. \tag{10.2}$$

Therefore, the horizon distance is setting the size of the observable universe. It is also interesting to consider another important aspect. If a causal message (an interacting signal) is emitted at some time t_{em}, then at any following time $t = t_{\text{em}} + \Delta t$, with $c\,\Delta t \ll R_H(t_{\text{em}}) \equiv c\,H^{-1}(t_{\text{em}})$, one simply has

$$\Delta\,d_{H,0}(t_{\text{em}}) = \int_{t_{\text{em}}}^{t_{\text{em}}+\Delta t} \frac{c\,dt}{a(t)} \simeq \frac{c\,\Delta t}{a_{\text{em}}} \ll \frac{c\,H^{-1}(t_{\text{em}})}{a_{\text{em}}} = R_H^{(0)}(t_{\text{em}}). \tag{10.3}$$

This means that regions that at any time are much smaller than the Hubble radius become causally connected within a period of time during which the expansion is negligible (i.e., the scale factor does not change significantly). Therefore, a period of time necessarily much shorter than the age of the Universe. This implies that such regions become causally connected at that epoch. For this reason, the Hubble radius R_H plays a very important role and can be regarded as an *instantaneous horizon*. This means that an event occurring at a certain point can influence what happens in another point at a distance below R_H.[1] Such length scales are said to be *sub-horizon* sized. Of course, if the length scale of a certain region becomes smaller than R_H at some past time $t_\star$ (it is said that it enters the horizon) then it is certainly smaller than $d_H(t_{\text{in}}, t_\star)$ at that time, if not earlier, and it will always remain smaller at any future time. This remains true even if it becomes *super-horizon* sized, i.e., bigger than R_H, at some time after $t_\star$ (it is said in this case that it exits the horizon). In this last case points of this region, though no longer in causal contact with each other at present, might still retain memory of the past age when they were causally connected. This possibility will play a crucial role in inflation.

In the case of models with an initial singularity, placing the origin of time at the singularity, the horizon distance can be written simply as

$$d_{H,0} = \int_0^{t_0} \frac{c\,dt}{a(t)} = c \int_0^1 \frac{da}{a^2\,H(a)}. \tag{10.4}$$

This quantity can be either finite or infinite and it can be infinite even if t_0 is finite. If it is infinite, this would mean that the horizon can be arbitrarily large and in principle the observations could encompass an arbitrarily large portion of the universe.

As an important example, let us consider again a one-fluid flat universe with $p = w\,\varepsilon$ and $w = \text{const} > -1$. In this case we have seen that the scale factor

---

[1] On the Earth, the usual terrestrial horizon plays the role of the Hubble radius, while the maximum causally connected region is the size of the maximum region visited by some explorer and it can be as large as the maximum distance on the Earth, $\pi\,R_\oplus$. An explorer interacts at a specific time with all points within the terrestrial horizon but the causally connected region can be much bigger, corresponding to the maximum distance that he has travelled within his life. This should illustrate how the Hubble radius should be regarded as an instantaneous horizon, while d_H as an integrated horizon.

evolution is described by Eq. (9.1). Inserting this expression into Eq. (10.4), one finds that the horizon distance for ever-decelerating models, with $w > -1/3$, is finite and given by

$$d_{H,0} = \frac{3(1+w)}{(1+3w)} \, c \, t_0 \, . \tag{10.5}$$

This means that in these models the observable universe has a finite size. All events outside the horizon are causally disconnected and cannot have (or have had) any influence on us.[2]

Using the expression Eq. (7.36) for the age of the universe, one can also express the horizon distance at present in terms of the Hubble radius at present $R_{H,0} \equiv c \, H_0^{-1}$, finding $d_{H,0} = 2 \, c \, H_0^{-1}/(1 + 3 \, w)$, showing that for a radiation universe ($w = 1/3$) this coincides with $R_{H,0}$, while for a matter universe ($w = 0$) it is twice $R_{H,0}$. This means that for standard fluids horizon distance and Hubble radius basically coincide, barring $\mathcal{O}(1)$ factors. This result is actually valid, more generally, at any time:

$$d_H(t) = \frac{2 \, c \, H^{-1}(t)}{1 + 3 \, w} \, . \tag{10.6}$$

We will be back on this point in Chapter 15.

10.2 TESTING MODELS WITH COSMOLOGICAL REDSHIFTS

In this section, we finally discuss how, by measuring the distance of an astronomical object and its redshift, one can reconstruct experimentally the expansion law $a(t)$ from the present time back to past times. This information can then be used to measure the cosmological parameters identifying the correct cosmological model.

As we have seen in the first two chapters, this was exactly the kind of procedure adopted by Hubble. However, from *Hubble's law*,

$$z \simeq \frac{H_0 \, d_L}{c} \, , \tag{10.7}$$

one can only determine the Hubble constant $H_0 = \dot{a}_0$, i.e., the expansion rate at the present time. This is because Hubble's law is an approximated law holding only at low redshifts. In order to gain information on the expansion law in the past, one has to be able to observe more distant objects, at higher redshift. In this way astronomical observations work effectively as a time machine.

Let us start from the calculation of the proper distance of a luminous source located in a point with comoving radial coordinate r_{em}, corresponding to an emission time t_{em}. In Chapter 6 we found the expression (6.22) that we write here again for convenience,

$$d_{pr,0}(r_{em}) = \int_{t_{em}}^{t_0} \frac{c \, dt}{a(t)} \, . \tag{10.8}$$

---

[2]For ever-accelerating models with $-1/3 > w > -1$ the horizon distance is infinite even though t_0 is finite. Can you explain why, both mathematically and physically? We will answer his question when we will discuss inflation in Chapter 15.

If one could measure directly $d_{\mathrm{pr},0}(r_{\mathrm{em}})$ on the left-hand side, this would place constraints on the cosmological parameters that determine $a(t)$ on the right-hand side. However, unfortunately, the proper distance is not a directly measurable quantity.

As we discussed in Chapter 2, what the astronomical observations typically measure is the luminosity distance defined by Eq. (2.16). In a static Euclidean universe proper distance and luminosity distance would coincide. However, in the expanding universe the two distances do not coincide since the geometry is in general curved and because the universe is expanding. We need then to understand the relation between luminosity distance and proper distance.

First of all we need to express the detected radiative flux F in terms of the comoving coordinate r_{em} of the emitting source. Remember that F is incident energy per unit time per unit area.

If the universe is curved, the area of the infinitesimal surface at distance $d_{\mathrm{pr}}(t_0)$ subtending a solid angle $d\Omega$, is not given by $d_{\mathrm{pr}}^2\, d\Omega$ but by $R_0^2\, r_{\mathrm{em}}^2\, d\Omega$, as it can be well understood using a two-dimensional analogy.[3] Therefore in a static curved universe, like for example in Einstein's static model, the radiative flux from a source with absolute luminosity L would be given by

$$F = \frac{L}{4\pi\, R_0^2\, r_{\mathrm{em}}^2}\,. \tag{10.9}$$

More generally, we have also to take into account two effects coming from the expansion of the universe. First, the energy of photons is redshifted, in a way that at the detection

$$E_0 = E_{\mathrm{em}}\, a_{\mathrm{em}} = \frac{E_{\mathrm{em}}}{(1+z)}\,, \tag{10.10}$$

and, second, the interval of time δt_0 between two photons at detection, is related to the interval of time δt_{em} at emission by[4]

$$\delta t_0 = \frac{\delta t_{\mathrm{em}}}{a_{\mathrm{em}}} = \delta t_{\mathrm{em}}\,(1+z)\,. \tag{10.11}$$

Because of this, the flux is further reduced by a factor $1+z$. Combining together these three effects, one finally finds

$$F = \frac{L}{4\pi\, R_0^2\, r_{\mathrm{em}}^2\,(1+z)^2}\,, \tag{10.12}$$

and from Eq. (2.16)

$$d_L(r_{\mathrm{em}}) = R_0\, r_{\mathrm{em}}\,(1+z)\,. \tag{10.13}$$

---

[3]In a two-dimensional analogy the size $d\ell$ of an infinitesimal object subtending an angle $d\theta$ with cylindrical coordinate r would be given by $d\ell = R_0\, r_{\mathrm{em}}\, d\theta$.

[4]We omit the demonstration of this last relation because it essentially coincides with the same demonstration made in Chapter 6 when we showed that the wavelength of an electromagnetic wave at the detection is related to the wavelength at the emission by Eq. (6.18).

At the same time, from the definition of proper distance Eq. (6.20), one finds:

$$d_{\text{pr},0}(r_{\text{em}}) = R_0 \arcsin(r_{\text{em}}), \quad \text{if } k = +1; \tag{10.14}$$

$$d_{\text{pr},0}(r_{\text{em}}) = R_0 \, r_{\text{em}} = \frac{d_L(r_{\text{em}})}{1+z}, \quad \text{if } k = 0; \tag{10.15}$$

$$d_{\text{pr},0}(r_{\text{em}}) = R_0 \arcsinh(r_{\text{em}}), \quad \text{if } k = -1. \tag{10.16}$$

In this way we are able to replace the proper distance with the luminosity distance in Eq. (10.8).

In order to simplify the discussion, let us focus on simpler *flat models* since we will see that this is the case supported by the observations. Therefore, Eq. (10.8) can be recast in terms of d_L simply as

$$d_L(r_{\text{em}}) = (1+z) \int_{t_{\text{em}}}^{t_0} \frac{c \, dt}{a(t)}. \tag{10.17}$$

We have now on the left-hand side a quantity that can be measured experimentally and on the right-hand side a function of the parameters of the model. For example, in the case of a Lemaitre model, a mixture of matter, radiation and a cosmological constant, the parameters can be chosen to be H_0, $\Omega_{R,0}$, $\Omega_{M,0}$ and $\Omega_{\Lambda,0}$.

In this way, the astronomical observations can be used to test cosmological models determining the cosmological parameters. The right-hand side of Eq. (10.17) can be manipulated in order to show in a more explicit way the dependence on the cosmological parameters. Moreover, we have to answer an interesting question: is Eq. (10.17) able to reproduce *Hubble's law*?

There are two ways to proceed: either one assumes a precise theoretical model and expresses (numerically or analytically) $a(t)$ as an exact function of the parameters of the model, or otherwise, without assuming any particular model, one can always, *model independently*, perform a Taylor expansion about $t = t_0$, obtaining

$$\frac{1}{a(t)} = \frac{1}{a_0} - \frac{\dot{a}}{a^2}\Big|_{t=t_0} (t - t_0) + \frac{1}{2}\left[2\frac{\dot{a}^2}{a^3} - \frac{\ddot{a}}{a^2}\right]_{t=t_0} (t - t_0)^2 + \mathcal{O}[(t_0 - t)^3]. \tag{10.18}$$

From this expansion, recalling that $a_0 = 1$ and $\dot{a}_0 = H_0$, one can then write

$$\frac{1}{a(t)} = 1 + H_0 (t_0 - t) + \left(1 + \frac{q_0}{2}\right) H_0^2 (t_0 - t)^2 + \mathcal{O}[(t_0 - t)^3], \tag{10.19}$$

where we introduced the dimensionless *deceleration parameter*

$$q(t) \equiv -\frac{\ddot{a}(t)}{a(t) \, H(t)^2} \Rightarrow q_0 = -\frac{\ddot{a}_0}{H_0^2}. \tag{10.20}$$

We can then easily obtain a Taylor expansion for the proper distance calculating the integral

$$d_{\text{pr},0}(t_{\text{em}}) = c \int_{t_{\text{em}}}^{t_0} \frac{dt}{a(t)} = c \, (t_0 - t_{\text{em}}) + \frac{1}{2} c \, H_0 \, (t_0 - t_{\text{em}})^2 + \mathcal{O}[(t_0 - t_{\text{em}})^3]. \tag{10.21}$$

The time difference $t_0 - t_{\rm em}$ is called *lookback time*. It is not a quantity that we can measure directly, but it can be expressed in terms of the redshift z. If we rewrite Taylor expansion Eq. (10.19) as

$$z = H_0 \left(t_0 - t_{\rm em}\right) + \left(1 + \frac{q_0}{2}\right) H_0^2 \left(t_0 - t_{\rm em}\right)^2 + \mathcal{O}[(t_0 - t_{\rm em})^3], \qquad (10.22)$$

this can then be easily inverted, yielding for the lookback time

$$t_0 - t_{\rm em} = H_0^{-1} \left[z - z^2 \left(1 + \frac{q_0}{2}\right)\right] + \mathcal{O}(z^3). \qquad (10.23)$$

If this relation is now plugged into the Taylor expansion for $d_{\rm pr,0}$, Eq. (10.21), we obtain

$$d_{\rm pr,0}(z) = c\, H_0^{-1} \left[z - z^2 \left(\frac{1 + q_0}{2}\right)\right] + \mathcal{O}(z^3). \qquad (10.24)$$

As a last step, we can then finally obtain an expression for the luminosity distance d_L as a function of the redshift z

$$d_L(z) = (1+z)\, d_{\rm pr,0}(z) = c\, H_0^{-1} \left[z + z^2 \left(\frac{1 - q_0}{2}\right)\right] + \mathcal{O}(z^3). \qquad (10.25)$$

This shows that Hubble's law is only valid at the first order in z, for $z \lesssim 0.1$. Considering that $c\, H_0^{-1} \simeq 4.5\,{\rm Gpc}$, it applies only for distances $d_L \lesssim 450\,{\rm Mpc}$. At higher distances, one has to take into account the z^2 correction term for $0.1 \lesssim z \lesssim 1$. Obviously, for $z \gtrsim 1$, the Taylor expansion breaks down and one has to go back to the exact expression Eq. (10.17). This, however, is model dependent and requires a specific knowledge of $a(t)$.

10.3 DECELERATION PARAMETER IN MULTI-FLUID MODELS

Let us consider a model with a mixture of fluids with $p_i = w_i\, \varepsilon_i$. The acceleration equation then reads

$$\frac{\ddot{a}}{a} = -\frac{4\,\pi\,G}{3\,c^2} \sum_i \varepsilon_i \left(1 + 3w_i\right). \qquad (10.26)$$

Remembering the definition of the critical energy density, Eq. (6.41), one obtains for the deceleration parameter

$$q(t) = \frac{1}{2} \sum_i \Omega_i(t) \left(1 + 3w_i\right), \qquad (10.27)$$

and, specifically at the present time,

$$q_0 = \frac{1}{2} \sum_i \Omega_{i,0} \left(1 + 3\,w_i\right). \qquad (10.28)$$

In the case of Lemaitre models with matter ($w_M = 0$), radiation ($w_R = 1/3$) and a cosmological constant ($w_\Lambda = -1$), this expression specialises into

$$q_0 = \Omega_{R,0} + \frac{1}{2}\Omega_{M,0} - \Omega_{\Lambda,0} \,. \tag{10.29}$$

If $\Omega_{\Lambda,0} \leq 0$, then necessarily $q_0 > 0$. On the other hand if $\Omega_{\Lambda,0} \geq 0$, then q_0 can be negative. This clearly confirms, in terms of q_0, that an accelerating expansion is possible only in the presence of a positive cosmological constant.

Notice that if $\Omega_{\Lambda,0} = 0$ and assuming $\Omega_{R,0} \ll \Omega_{M,0}$ (as confirmed by other cosmological observations), then the model depends only on one parameter $\Omega_0 \simeq \Omega_{M,0}$. In this case, a measurement of q_0 gives a full characterisation of the model. What kind of information on the cosmological parameters do we obtain from cosmological observations?

10.4 THE EXPANSION OF THE UNIVERSE IS ACCELERATING AT PRESENT

As we discussed, in 2001 the Hubble Space Telescope managed to measure H_0 with quite a good accuracy by measuring distances below $\sim 400\,\mathrm{Mpc}$ (corresponding to redshifts below ~ 0.1). At these distances the effect of the deceleration parameter is absent and therefore one can obtain a measurement of H_0 from the slope of the approximate linear relation between luminosity distance and redshift (see right panel of Fig. 2.1). The precision of this measurement was mainly due to a Cepheid calibration of different secondary methods including supernovae type Ia (SNIa).

In 1998 two research teams, the *High-z Supernova Search Team* [1] and *Supernovae Cosmology Project* [2], released the results from searches of SNIa from very distant galaxies (up to redshifts $z \sim 1$). The two collaborations found that a plot of the *distance modulus*, defined as

$$m - M \equiv 5 \log\left(\frac{d_L}{1\,\mathrm{Mpc}}\right) + 25 \,, \tag{10.30}$$

basically the logarithm of the luminosity distance, versus the redshift is well fitted within Lemaitre models. In this case, the supernovae data translate into an allowed region in the plane $\Omega_{M,0} - \Omega_{\Lambda,0}$ corresponding approximately to

$$\Omega_{\Lambda,0} \simeq 1.5\,\Omega_{M,0} + 0.25 \,. \tag{10.31}$$

This result indicates that within this model a non-vanishing positive cosmological constant is necessary in order to explain the data. Moreover, using the relation Eq. (10.29) for the deceleration parameter, one finds

$$q_0 = -\Omega_{M,0} - 0.25 < 0 \,. \tag{10.32}$$

Therefore, the SNIa data indicate that the universe expansion is accelerating at present. New data released at the end of 2003 [3] and at the beginning of 2004 [4] confirmed and even strengthened these conclusions with considerably reduced

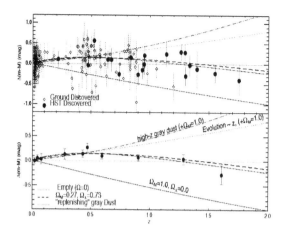
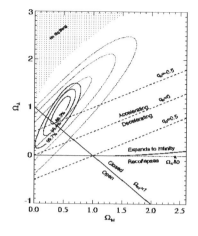

FIGURE 10.1: Results from high redshifts type Ia supernova searches [4]. Left panel: Plot of measured residuals of the distance modulus, compared to the predictions from an empty model ($q_0 = 0$), for about 200 type Ia supernovae versus redshift. Right panel: Constraints on the plane Ω_M - Ω_Λ coming from fitting the data within Lemaitre models.

experimental errors. In Fig. 10.1 we show the results from [4]: in the left panel the plot of the distance modulus versus redshift and in the right panel the constraints in the $\Omega_{M,0} - \Omega_{\Lambda,0}$ plane. As we will discuss, today further cosmological observations confirm independently this astonishing result of an accelerated expansion at present.

10.5   EXPANSION IN THE ΛCDM MODEL

In addition to the SNIa data, constraining approximately the difference between $\Omega_{\Lambda,0}$ and $\Omega_{M,0}$, from CMB anisotropies observations, [5] one finds a strong indication that the universe is approximately flat, implying, within Lemaitre models,

$$\Omega_0 \simeq \Omega_{M,0} + \Omega_{\Lambda,0} \simeq 1 \,. \tag{10.33}$$

Combining this constraint with Eq. (10.31), one finds straightforwardly $\Omega_{M,0} \simeq 0.3$ and $\Omega_{\Lambda,0} \simeq 0.7$. [6] This result is nicely confirmed by independent measurements from baryon acoustic oscillations (BAO) in galaxy distribution observations that approximately measure $\Omega_{M,0} \sim 0.3$. This beautiful agreement of three completely independent cosmological observations, shown in Fig. 10.2, is usually referred to as *cosmological concordance* and provides strong support to a description of the expansion of the universe in terms of a flat Lemaitre model containing both matter and a cosmological constant-like fluid. This is a first very important feature

---

[5]We will discuss this result in Chapter 12.

[6]Using these values, from the relation Eq. (10.32), one finds a value for the deceleration parameter at present given by

$$q_0 \simeq -0.55 \,. \tag{10.34}$$

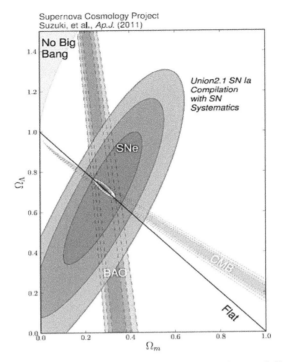

FIGURE 10.2: Cosmic concordance (from [5]).

of the so-called ΛCDM model, the cosmological model that currently explains all cosmological observations. As we will see, there are other important properties of this model. We have already discussed in Section 8.4.1 the stages of the expansion within a flat Lemaitre model.

We can now specify the values of the different quantities that describe the onset of the different regimes. For example, we can calculate the value of the scale factor at the time of equality between matter and Λ energy densities. This is given by

$$a_{\text{eq}}^{M\Lambda} = \left(\frac{\Omega_{M,0}}{\Omega_{\Lambda,0}}\right)^{\frac{1}{3}} \simeq 0.75 \, . \tag{10.35}$$

From Eq. (9.26) and using Eq. (2.30) for H_0^{-1}, the Hubble time, we can calculate numerically $t_{\text{eq}}^{M\Lambda}$, finding

$$t_{\text{eq}}^{M\Lambda} \simeq (10.2 \pm 0.2)\,\text{Gyr}\,, \tag{10.36}$$

where we used $\Omega_{\Lambda,0} = 0.692 \pm 0.012$ and $\Omega_{M,0} = 0.308 \pm 0.012$, more precise values found from CMB anisotropies observations. From Eq. (9.21) we also find

$$t_0\,H_0 = f(\Omega_{\Lambda,0} \simeq 0.7) \simeq 0.96 \quad \Rightarrow \quad t_0 \simeq 13.8\,\text{Gyr}\,. \tag{10.37}$$

Notice that having $t_0\,H_0 \simeq 1$ is just a coincidence in the ΛCDM model. It is also a (related) coincidence that the values of $\Omega_{M,0}$ and $\Omega_{\Lambda,0}$ are comparable with

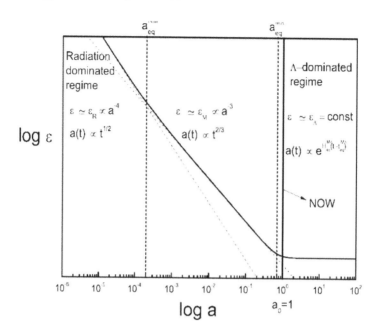

FIGURE 10.3: Evolution of the energy density in the ΛCDM model (without inflationary stage). There are three well-distinguished regimes characterised by the dominance of one of the three fluids: matter, radiation and the cosmological constant in chronological order. It has to be noticed how the onset of the Λ-dominated regime, characterised by an accelerated expansion, started only very recently: it is sometimes informally referred to as the *why now problem*.

each other and that the cosmological constant contribution started to dominate the total energy density of the universe relatively recently (in cosmological time!): only 4 billion years ago. This coincidence becomes clear when the cosmological history is represented in a logarithmic plot shown in Fig. 10.3. We can also finally calculate the contribution to $\Omega_{R,0}$ from the CMB, verifying that it is negligible compared to $\Omega_{M,0}$ and $\Omega_{\Lambda,0}$. Using Eq. (6.42) for the critical energy density at present and Eq. (2.36) for the energy density of CMB photons at present, one finds

$$\Omega_{\gamma,0}^{CMB} \simeq 0.25 \times 10^{-4} \, h^{-2} \simeq 0.537 \times 10^{-4} \,. \tag{10.38}$$

Since the energy density of radiation from astronomical objects contributes less than 10% compared to the energy density of CMB radiation, then $\Omega_{\gamma,0}^{CMB}$ gives, with good approximation, the contribution of photons to $\Omega_{R,0}$. However, as we will discuss in Chapter 13, there is another comparable contribution coming from the most elusive known elementary particles: neutrinos. For the time being we neglect this contribution approximating $\Omega_{R,0} \simeq \Omega_{\gamma,0}^{CMB}$. However, it should be clear that even accounting for neutrinos, the result $\Omega_{R,0} \ll \Omega_{M,0}, \Omega_{\Lambda,0}$ remains valid anyway.

From the value of $\Omega_{R,0}$ we can also estimate the value of the scale factor and of the redshift at matter-radiation equality time, finding

$$a_{\text{eq}} = \frac{\Omega_{R,0}}{\Omega_{M,0}} \simeq 1.73 \times 10^{-4} \quad \Rightarrow \quad z_{\text{eq}} = \frac{1}{a_{\text{eq}}} - 1 \simeq 6000\,. \tag{10.39}$$

We can also calculate the matter-radiation equality time finding[7] $t_{\text{eq}} \simeq 23,000\,\text{yr}$. As we said, we will have to inspect whether and how a potential contribution from neutrinos can change these numbers.

In conclusion, the observations of the cosmological redshifts indicate that the universe expansion is described by three regimes (see Fig. 10.3). The radiation-dominated regime describes the initial stage of the universe and is followed by a matter-dominated regime that turned, only 4 billion years ago, into a Λ-dominated regime described by an (exponential) de Sitter expansion. There is only one stage still missing in this picture and that represents another important ingredient of the ΛCDM model: the inflationary stage. This stage precedes the radiation-dominated regime and describes the history of so-called *very early universe*. We will discuss inflation in Chapter 15.

EXERCISES

Exercise 10.1 *Consider a flat universe containing one fluid with equation of state $p/\varepsilon = w = $ const. Derive the proper distance at present $d_{\text{pr},0}$ and the luminosity distance $d_L(z)$ as a function of z in terms of w.*

Exercise 10.2 *Let us reconsider the Milne–McCrea model in more detail as to whether it can successfully reproduce the observed cosmological redshifts. Replace the non-relativistic expression for the Doppler effects with the relativistic one and calculate the predicted value for the deceleration parameter, showing that this is in disagreement with the measured value $q_0 \simeq -0.55$.*

Exercise 10.3 *Using the estimated value for $\Omega_{R,0} \simeq \Omega_{\gamma,0}^{CMB} \simeq 0.537 \times 10^{-4}$, show that the the matter-radiation equality time $t_{\text{eq}} \simeq 23,000\,\text{yr}$.*

Exercise 10.4 *Given the measured values for the parameters of the ΛCDM model, how long will it take for the scale factor to increase a factor of 4 from now?*

Exercise 10.5 *In Fig. 10.2 there is an excluded region labelled 'No Big Bang'. Explain why it is excluded and derive the corresponding constraint in the plane $\Omega_M - \Omega_L$.*

BIBLIOGRAPHY

[1] A. G. Riess et al. [Supernova Search Team], *Observational evidence from supernovae for an accelerating universe and a cosmological constant*, Astron. J. **116** (1998) 1009 [astro-ph/9805201].

---

[7]See Exercise 10.3.

[2] S. Perlmutter et al. [Supernova Cosmology Project Collaboration], Astrophys. J. **517** (1999) 565 [astro-ph/9812133].

[3] R. A. Knop et al. [Supernova Cosmology Project Collaboration], *New constraints on Omega(M), Omega(lambda), and w from an independent set of eleven high-redshift supernovae observed with HST*, Astrophys. J. **598** (2003) 102 doi:10.1086/378560 [astro-ph/0309368].

[4] A. G. Riess et al. [Supernova Search Team], *Type Ia supernova discoveries at z ¿ 1 from the Hubble Space Telescope: Evidence for past deceleration and constraints on dark energy evolution*, Astrophys. J. **607** (2004) 665 [astro-ph/0402512].

[5] R. Amanullah et al., *Spectra and light curves of six type Ia supernovae at 0.511 < z < 1.12 and the Union2 compilation*, Astrophys. J. **716** (2010) 712 [arXiv:1004.1711 [astro-ph.CO]].

CHAPTER 11

Matter

A s discussed in the last chapter, the cosmological observations point to a model where most of the energy density of the universe, about 69%, is in the form of a cosmological constant or of some fluid that closely mimics it, and matter contributes approximately to the remaining 31% fraction. In this chapter we finally take a closer look at the matter fluid, examining the contribution from different astrophysical components, discussing how it is distributed in the universe and confirming in an independent direct way the result $\Omega_{M,0} \simeq 0.31$. This establishes an important *cosmological concordance* among different independent observations in the determination of the cosmological energy budget. At the same time, we will see how an intriguing cosmological puzzle starts to emerge: most of the matter does not seem to be associable to any ordinary component. This unidentified form of matter is the dark matter of the universe. In the next chapters we will see that further arguments confirm the existence of dark matter and show its non-baryonic nature. Finally, in Chapter 17 we will discuss some of the most attractive proposals on the nature of this mysterious fluid that has so far escaped all attempts to determine it.

11.1 MATTER IN STARS

Let us first of all try to understand where the ordinary matter is mainly located in the universe. Stars are the first place to search for matter in the universe. However, matter in stars can account only for a tiny fraction of the total matter energy. The total luminosity density of stars in the so-called *B-band*[1] is given by

$$j_{\star,B} \simeq 1.2 \times 10^8 \, L_{\odot,B} \, \text{Mpc}^{-3} \,. \tag{11.1}$$

On the other hand, the *mass-to-light ratio* for the stars is given approximately, for light in the B-band by,

$$\left\langle \frac{M}{L_B} \right\rangle_\star \simeq 4 \, \frac{M_\odot}{L_{\odot,B}} \,. \tag{11.2}$$

---

[1] It is blue light, with a wavelength range $4.0 \times 10^{-7} \, \text{m} < \lambda < 4.9 \times 10^{-7} \, \text{m}$. We will denote this band with the subscript B that, notice, we also adopted to indicate the baryonic contribution.

Therefore, one finds that the energy density stored in the mass of the stars at present, is approximately given by

$$\varepsilon_{\star,0} \simeq \left\langle \frac{M}{L_B} \right\rangle_\star \, j_{\star,B} \, c^2 \simeq 5 \times 10^8 \, M_\odot \, c^2 \, \mathrm{Mpc}^{-3} \,. \tag{11.3}$$

The critical energy density (see Eq. (6.42)) can also be expressed in solar units using Eqs. (2.1), (2.4) and (2.7), finding

$$\varepsilon_{c,0} \simeq 1.4 \times 10^{11} \, M_\odot \, c^2 \, \mathrm{Mpc}^{-3} \,. \tag{11.4}$$

In conclusion, we find that the contribution from stars to the matter energy density parameter is given by

$$\Omega_{\star,0} = \frac{\varepsilon_{\star,0}}{\varepsilon_{c,0}} \simeq 0.004 \,. \tag{11.5}$$

It is then clear that most of the matter has to reside elsewhere, not in stars.

11.2 MATTER IN HOT LOW-DENSITY GAS

From X-ray astronomy, we know that hot low density gas fills the space between galaxies in clusters of galaxies and emits X-ray photons with an average energy of about $10\,\mathrm{keV}$. It is estimated that the total energy stored in the mass of this intra-cluster gas is about seven times the contribution from stars and therefore

$$\Omega_{\mathrm{gas},0} \simeq 0.03 \,. \tag{11.6}$$

It is clear that this contribution is still insufficient to account for the value of $\Omega_{M,0}$ found in the last chapter, combining SNIa and CMB data.

11.3 MATTER IN GALAXIES

If stellar orbits are approximated with a circular motion, then the acceleration acting on a star at a distance R from the galactic centre is simply related to the speed of the star by

$$a = \frac{v^2}{R} \tag{11.7}$$

and is directed toward the centre of the galaxy. At the same time, from Newton's law, we can easily relate the gravitational acceleration to the mass $M_G(R)$ of the galaxy within a distance R from the galactic centre, writing

$$a = \frac{G \, M_G(R)}{R^2} \,. \tag{11.8}$$

In this way, we can estimate the velocity of the star as

$$v(R) = \sqrt{\frac{G \, M_G(R)}{R}} \,. \tag{11.9}$$

Experimentally, from the surface brightness of the galactic disk, the density of the stars within a distance R is found to be well described by an exponential behaviour that can be written explicitly as

$$\rho_\star(R) = \rho_\star(0) \, \exp\left(-\frac{R}{R_s}\right) , \qquad (11.10)$$

where $R_s = \mathcal{O}(\text{kpc})$.[2] Therefore, most of the visible mass of a galaxy is concentrated within a few kiloparsecs from its centre implying $M_G(R) \simeq \text{const}$ for $R \gtrsim 2\,R_s$.

The velocity of external objects should then drop down following the Keplerian behaviour,

$$v(R \gtrsim 2\,R_s) \propto \frac{1}{\sqrt{R}} , \qquad (11.11)$$

the same describing planetary velocities in our solar system within a circular orbit approximation. However, data shows that even at distances as large as $\sim 100\,\text{kpc}$, the galactic velocities remain approximately constant from the centre [1, 2]. Inverting Eq. (11.9), this result implies

$$M_G(R) \simeq \frac{v^2\,R}{G} \simeq 10^{12}\,M_\odot \left(\frac{v}{220\,\text{km } s^{-1}}\right)^2 \left(\frac{R}{100\,\text{kpc}}\right) \qquad (R \lesssim R_\text{halo}), \quad (11.12)$$

where R_halo is the size of the so-called galactic halo. Given that the luminosity of a galaxy in the B-band is about

$$L_{G,B} \simeq 2 \times 10^{10}\,L_{\odot,B} , \qquad (11.13)$$

one finds a *mass-to-light ratio* in galaxies

$$\left\langle \frac{M}{L_B} \right\rangle_G \simeq 50\,\frac{M_\odot}{L_{\odot,B}} \left(\frac{R_\text{halo}}{100\,\text{kpc}}\right) \left(\frac{v}{220\,\text{km } s^{-1}}\right)^2 . \qquad (11.14)$$

A comparison with Eqs. (11.2) and (11.5) immediately yields, for typical velocities $v \simeq 220\,\text{km } s^{-1}$, an estimated value of $\Omega_{M,0}$ in galaxies from galactic rotation curves,

$$\Omega_{M,0}^G \simeq 0.05 \left(\frac{R_\text{halo}}{100\,\text{kpc}}\right) . \qquad (11.15)$$

The most recent estimations indicate for the radii of typical galactic halos, such as in the case of Milky Way and M31, $R_\text{halo} \sim 300\,\text{kpc}$ [3] and in this case one obtains

$$\Omega_{M,0}^G \simeq 0.15 \gg \Omega_{\star,0} + \Omega_{\text{gas},0} . \qquad (11.16)$$

This result shows that galactic rotation curves provide evidence for the existence of a *dark matter component*.

Notice, however, that $\Omega_{M,0}^G$ still underestimates the total $\Omega_{M,0}$. This is because it is not sensitive to a possible smooth component of dark matter filling all space among galaxies in clusters. The effects of this additional intergalactic dark matter component can be inferred from the motion of the galaxies in the clusters of galaxies.

---

[2]For example for our galaxy $R_s \sim 4\,\text{kpc}$.

11.4 MATTER IN CLUSTERS OF GALAXIES

Additional strong evidence for the existence of matter not in the form of stars and hot gas comes from the dynamics of galaxies within clusters of galaxies. The Newtonian dynamics obeys the *virial theorem*, stating that for a self-gravitating system in steady state one has (see Exercise 11.1)

$$K = -\frac{W}{2}, \tag{11.17}$$

where K is the kinetic energy and W is the potential energy. The kinetic energy can be expressed as

$$K = \frac{1}{2} M \langle v^2 \rangle, \tag{11.18}$$

where M is the total mass of the cluster and $\langle v^2 \rangle$ is the mean squared velocity. The potential energy can be written as

$$W = -\alpha \frac{G M^2}{r_{\rm h}}, \tag{11.19}$$

where α is a $\mathcal{O}(1)$ numerical factor and $r_{\rm h}$ is the *half mass radius of the cluster*, i.e., the radius of a sphere centred on the cluster's centre of mass and containing a mass $M/2$. Therefore, from the virial theorem, we can write

$$M \langle v^2 \rangle = \alpha \frac{G M^2}{r_{\rm h}}, \tag{11.20}$$

obtaining

$$M = \frac{r_{\rm h} \langle v^2 \rangle}{\alpha G}. \tag{11.21}$$

Notice that this equation is a sort of generalisation of Eq. (11.12) that has now been obtained for self-gravitating systems with much more complicated and general Newtonian dynamics. The values of the involved parameters, α, $r_{\rm h}$ and $\langle v^2 \rangle$ depend on the specific considered cluster of galaxies and therefore have to be somehow extracted from observations of individual clusters.

This has been possible with good precision for the Coma cluster,[3] where

$$(r_{\rm h})_{\rm Coma} \simeq 1.5\,{\rm Mpc}\,, \quad \alpha_{\rm Coma} \simeq 0.4\,, \quad \langle v^2 \rangle_{\rm Coma} \simeq (1.500\,{\rm km}\ s^{-1})^2\,. \tag{11.22}$$

Plugging these numbers into Eq. (11.21) and using the value of G in solar units Eq. (2.12), one finds

$$M_{\rm Coma} \simeq 2 \times 10^{15}\,M_{\odot}\,. \tag{11.23}$$

---

[3]This is the cluster originally studied by Fritz Zwicky that provided the first hint of the existence of dark matter from clusters of galaxies [4]. A partial image of the Coma cluster is shown in the front cover page.

Combining this with the luminosity of the Coma cluster

$$L_{\text{Coma},B} \simeq 8 \times 10^{12} \, L_{\odot,B} \, , \tag{11.24}$$

one finds for the *mass-to-light ratio* of the Coma cluster [5]

$$\left\langle \frac{M}{L_B} \right\rangle_{\text{Coma}} \simeq 250 \, \frac{M_\odot}{L_\odot} \, . \tag{11.25}$$

Similar results are found for other clusters of galaxies. A study of 29,000 objects from the 2dF galaxy redshift survey catalogue finds [6]

$$\left\langle \frac{M}{L_B} \right\rangle_{\text{clusters}} \simeq (300 \pm 17) \, \frac{M_\odot}{L_\odot} \, . \tag{11.26}$$

Comparing again with Eqs. (11.2) and (11.5), this result for the mass-to-light ratio gives

$$\Omega_{\text{M},0}^{\text{clusters}} = (0.26 \pm 0.03) \, . \tag{11.27}$$

Notice that this is still lower than the value $\Omega_{M,0} \simeq 0.31$ inferred by combining the information from the cosmological redshifts of the supernovae with the flatness of the universe from CMB. This gap can be explained by a smooth component of dark matter among clusters of galaxies, in the so-called *voids*. Therefore, the indirect derivation of $\Omega_{M,0}$ is now confirmed by a more direct independent determination.

However, there is clearly still the problem to understand the difference between $\Omega_{M,0} \simeq 0.31$ and $\Omega_{\star,0} + \Omega_{\text{gas},0} \simeq 0.034$, suggesting the presence of a *dark matter* component[4] ultimately necessary to explain how stars in galaxies and galaxies in clusters can be bound with the high velocities we observe. The same can be said for the hot intra-cluster gas that remains bound despite its high velocity inferred by its mean temperature. We can therefore fairly summarise the results on the gravitational evidence for the existence of dark matter, saying that this acts as a sort of *cosmic glue* necessary to bind the stars in galaxies and the galaxies themselves in the clusters of galaxies at high measured speeds.

11.5 MASSIVE ASTROPHYSICAL COMPACT HALO OBJECTS (MACHOs)

A minimal astrophysical explanation of dark matter is that it is made of faint astrophysical objects. Ordinary matter is made basically of nucleons (protons and

---

[4]The name first appeared in 1906 in a work by Henri Poincaré in French, *matière obscure*, titled *The Milky Way and the theory of gases*. Here, in response to the idea of Lord Kelvin who proposed two years earlier that the presence of *dark bodies* could explain the measured stellar velocities in our Galaxy, he actually was showing that it was not possible to draw conclusions on the basis of existing data. The following works by the Dutch astronomers Jacobus Kapteyn (1922) and Jan Oort (1932) supported the idea of dark matter for a consistent explanation of stellar motion in our galaxy [7]. However, the first work that put the idea of dark matter within a cosmological perspective and gave it more serious consideration, was the 1933 paper by the Swiss astronomer Fritz Zwicky [4] on the study of the dynamics of clusters of galaxies in German where he was talking of *dunkle materie*. In addition he also even referred to it as *kalte*, i.e., cold. [8]

neutrons) and electrons. Because of the universe electric neutrality, the number density of electrons is equal to the number density of protons. Since $m_e \ll m_p$, the dominant contribution to the matter energy density at present is in the form of nucleons and, since nucleons are baryons, this is usually referred to as *baryonic matter* and its contribution to the energy density parameter is indicated with $\Omega_{B,0}$.

There is, in addition, another known component of matter that cannot strictly speaking be included in the baryonic budget: black holes. Therefore, a possibility is that the dark matter component might be explained in terms of a faint baryonic matter component, such as brown dwarfs and giant planets, jointly with a population of black holes. These are collectively referred to as *massive astrophysical compact halo objects* (MACHOs) [9] as opposed to popular dark matter WIMPs that we will discuss in Section 11.4.5.

There are many ways to place constraints on the MACHO abundance in the galactic halo. The most stringent ones are based on gravitational microlensing, first predicted by Einstein in 1936 [10]. If a massive body, like a MACHO, intervenes between the observer and a background source star in the halo, it causes a temporary increase of the source brightness.[5] It was suggested that the most convenient set of sources to monitor are stars in Magellanic clouds, with events lasting between a few hours and years, with the distribution peaking at about two months [11, 9]. Different surveys have discovered a few events in the halo of our galaxy. Their contribution to dark matter can nicely explain the baryonic dark matter contribution that is needed to fill the gap between the sum of the stars and hot gas contribution, $\Omega_{\star,0} + \Omega_{gas,0} \simeq 0.034$, and the baryonic contribution, $\Omega_{B,0} \simeq 0.05$, measured by CMB and Big Bang nucleosynthesis. However, they still cannot fill the much bigger gap with the total matter contribution $\Omega_{M,0} \simeq 0.31$.[6] The difference $\Omega_{DM,0} \equiv \Omega_{M,0} - \Omega_{B,0} \sim 0.26$ has to be ascribed to some mysterious form of non-baryonic component, or to some new physical process able to mimic it, commonly simply referred to as *dark matter*. The dark matter puzzle is one of the most intriguing mysteries in modern physics, since its solution seems to point to an extension of our current knowledge of fundamental physics, showing the importance of cosmology in guiding our search for physics beyond general relativity and standard model. This conclusion is further strengthened by the necessity that dark matter had to exist much earlier than stars and galaxies formed, playing a crucial role in galaxy

---

[5]In gravitational microlensing, as opposed to strong and weak lensing, the shape of a source object is not deformed by the gravitational lens, but its brightness is temporarily changed.

[6]The EROS-2 survey has monitored 33×10^6 stars in the Magellanic clouds over a period of about 7 years from July 1996 till February 2003, searching for microlensing events caused by MACHOs in the halo of the Milky Way galaxy [12]. Only one candidate event was found, whereas about 39 events were expected for MACHOs with masses $\sim 0.4 \, M_\odot$. This imposes a stringent upper bound on the contribution of MACHOs, in the form of giant planets and faint stars (in the mass range $10^{-6} \, M_\odot < M < 1 \, M_\odot$, to the halo mass of less than 8%. More generally, the EROS-2 survey excludes MACHOs as main components of dark matter in the range $0.6 \times 10^{-7} \, M_\odot < M < 15 \, M_\odot$. The OGLE-III survey has confirmed and strengthened these results [13]. This constraint would still leave open the option of dark matter made of supermassive black holes with masses above $15 \, M_\odot$. We will come back to this possibility in Section 11.4.3.

formation and more generally in the evolution of perturbations. The production of dark matter has then necessarily to take place during the early universe epoch, that we are now going to discuss in the second part.

EXERCISES

Exercise 11.1 *Derive the virial theorem for a more general system whose potential energy is a homogeneous function of coordinates of degree k, i.e., such that* $W(\alpha \vec{r}_1, \alpha \vec{r}_2, \ldots, \alpha \vec{r}_N) = \alpha^k\, W(\alpha\, \vec{r}_1, \alpha\, \vec{r}_2, \ldots, \alpha\, \vec{r}_N)$.

Exercise 11.2 *Derive (order-of-magnitude wise) the numerical expression Eq. (11.12) for the mass of a galaxy within a distance R from the centre.*

BIBLIOGRAPHY

[1] K. C. Freeman, *On the disks of spiral and S0 galaxies*, Astrophys. J. **160** (1970) 811.

[2] V. C. Rubin and W. K. Ford, Jr., *Rotation of the Andromeda nebula from a spectroscopic survey of emission regions*, Astrophys. J. **159** (1970) 379.

[3] A. Klypin, H. Zhao and R. S. Somerville, *ΛCDM-based models for the Milky Way and M31 I: dynamical models*, Astrophys. J. **573** (2002) 597.

[4] F. Zwicky, *Die Rotverschiebung von extragalaktischen Nebeln*, Helv. Phys. Acta **6** (1933) 110 [Gen. Rel. Grav. **41** (2009) 207].

[5] E. L. Lokas and G. A. Mamon, *Dark matter distribution in the Coma cluster from galaxy kinematics: Breaking the mass - anisotropy degeneracy*, Mon. Not. Roy. Astron. Soc. **343** (2003) 401 [astro-ph/0302461].

[6] V. R. Eke et al. [2dFGRS Team], *Galaxy groups in the 2dFGRS: The Luminous content of the groups*, Mon. Not. Roy. Astron. Soc. **355** (2004) 769.

[7] J. C. Kapteyn, *First attempt at a theory of the arrangement and motion of the sidereal system*, Astrophys. J. **55** (1922) 302; J.H. Oort, *The force exerted by the stellar system in the direction perpendicular to the galactic plane and some related problems*, Bull. Astron. Inst. Netherlands **6** (1932) 249.

[8] G. Bertone and D. Hooper, *A history of dark matter*, [arXiv:1605.04909].

[9] K. Griest, *Galactic microlensing as a method of detecting massive compact halo objects*, Astrophys. J. **366** (1991) 412.

[10] A. Einstein, Science **84** (1936) 506.

[11] B. Paczynski, *Gravitational microlensing by the galactic halo*, Astrophys. J. **304** (1986) 1.

[12] P. Tisserand et al. [EROS-2 Collaboration], *Limits on the MACHO content of the galactic halo from the EROS-2 Survey of the Magellanic clouds*, Astron. Astrophys. **469** (2007) 387 [astro-ph/0607207].

[13] L. Wyrzykowski et al., *The OGLE view of microlensing towards the Magellanic clouds. IV. OGLE-III SMC data and final conclusions on MACHOs*, Mon. Not. Roy. Astron. Soc. **416** (2011) 2949 [arXiv:1106.2925 [astro-ph.GA]].

II

The early universe

CHAPTER 12

The cosmic microwave background

I n this chapter we discuss how within Big Bang cosmological models it is possible to explain in a simple way the existence of the CMB radiation and its thermal equilibrium spectrum. At the same time we discuss how from the study of CMB temperature anisotropies it is possible to extract a stunningly rich amount of information on the cosmological parameters. In particular one of the main results is that Ω_0 is close to unity within one percent error, implying that the geometry of our observable universe can be approximated to be flat with very good precision.

12.1 OBSERVATIONAL RESULTS

There are three main observational results from the analysis of CMB data.

1. **Thermal spectrum**

   The CMB spectrum is incredibly well described by a thermal equilibrium spectrum (*Planckian spectrum*) [1] with an average temperature $T_0 = (2.7255 \pm 0.0006)\,K$ [2].

   The deviation of the observed intensity, $d\varepsilon_{\gamma,0}/d\nu$, from an exact Planckian spectrum is smaller than 10^{-4} [1].

   Consequently the CMB energy density and photon number density are very well approximated by the thermal equilibrium expressions, Eq. (2.34) and Eq. (2.36) respectively.

2. **Dipole anisotropy**

   The temperature of the CMB spectrum exhibits a dipole anisotropy $\Delta T/T \sim 10^{-3}$, perfectly explained by the Doppler effect induced by Earth's peculiar velocity, i.e., the motion of the Earth with respect to the comoving system with a velocity $v \simeq 370\,\mathrm{km\,s^{-1}}$ [3]. This velocity is the result of the composition of different motions (Earth's annual revolution around the Sun, solar system motion around the galactic centre, galactic motion in the Local Group, the infall of the Local Group toward the Hydra-Centaurus super-cluster) and the observed annual modulation due to the Earth's revolution provides a strong confirmation.

3. **Primary anisotropies**

For a generic point in the sky with angular coordinates θ and ϕ (the usual angular spherical coordinates), the temperature of the CMB can always be decomposed as the sum of the average value plus a fluctuation, explicitly

$$T_0(\theta, \phi) = \langle T_0 \rangle + \Delta T(\theta, \phi) \,. \tag{12.1}$$

As we will see, the fluctuation can be further decomposed in the sum of different multipole contributions. Once the dipole anisotropy is subtracted, the temperature fluctuation is given by higher multipoles primary anisotropies originating in the early universe. These are found to be much smaller than the dipole anisotropy, $\Delta T/T \sim 10^{-5}$. Primary anisotropies have been discovered by the COBE satellite [4] that produced the first map of the sky of CMB primary anisotropies. Though the poor resolution ($\Delta\theta \sim 7°$) did not allow to resolve their structure, COBE could measure their amplitude at large scales. The WMAP satellite obtained a new map of the sky of CMB temperature anisotropies with much higher resolution ($\Delta\theta \sim 10'$) [5] that allowed to resolve the structure and detect the so-called *acoustic peaks* in the power spectrum predicted theoretically much earlier [6]. Finally the *Planck* satellite has obtained the highest resolution map so far ($\Delta\theta \sim 3'$) [7]. A comparison of the maps obtained by COBE and *Planck* is shown in Fig. 12.1.

12.2 CMB AS A FOSSIL RADIATION

An important point is that the thermal equilibrium distribution of relic photons that we observe at present is preserved by the expansion (both in past and in future). Let us examine this important result in more detail.

12.2.1 Evolution of photon distribution during the expansion

Relic photons freely propagate and the only effect that one has to take into account is the momentum redshift, so that a value of the momentum at present, that we indicate with $|\vec{p}|_0$, corresponds to a past value

$$|\vec{p}|(t) = \frac{|\vec{p}|_0}{a(t)} \,. \tag{12.2}$$

Since photons are bosons and since their spectrum is very well approximated by a thermal equilibrium spectrum, at the present time the distribution function of relic photons is well approximated by a Bose–Einstein distribution with temperature T_0, so that we can write

$$f_{\gamma,0}(\vec{p}) = \frac{1}{e^{c\,|\vec{p}|/T_0} - 1} \equiv f_{BE}(\vec{p}, T_0) \,, \tag{12.3}$$

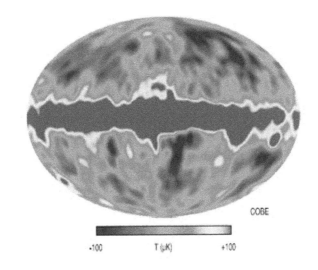

COBE

-100   T (μK)   +100

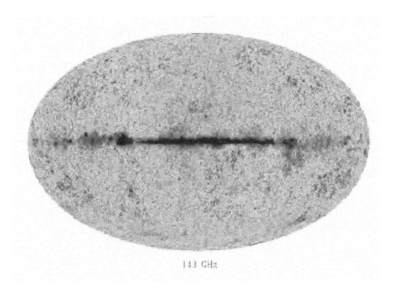

143 GHz

FIGURE 12.1: Sky maps of the CMB anisotropies obtained by the COBE [4] (top panel), and *Planck* [7] (bottom panel) satellites.

From Eq. (12.2), one has that photons with momentum $\vec{p}$ at some past time t, would have today a momentum $a(t)\,\vec{p}$, so that the distribution function at t can be written as[1]

$$f_\gamma(\vec{p}, t) = f_{\gamma,0}(a(t)\,\vec{p}) = \frac{1}{e^{a(t)|\vec{p}|\,c/T_0} - 1} \equiv f_{BE}(\vec{p}, T(t))\,. \qquad (12.4)$$

This distribution is still a (Bose–Einstein) thermal equilibrium distribution but now with a temperature

$$T(t) = \frac{T_0}{a(t)}\,. \qquad (12.5)$$

Therefore, the redshift of momenta translates into a dependence $T(t) \propto a(t)^{-1}$. This means that the CMB thermal equilibrium distribution that we observe at the present time can be explained assuming that the initial value of the scale factor was sufficiently small to guarantee the occurrence of an early hot stage in the history of the universe, when temperature was high enough to enforce thermal equilibrium. However, high temperature is not a sufficient condition to establish thermal equilibrium. The existence of processes able to *thermalise* photons, i.e., able to redistribute efficiently energy and momentum among quantum states of photons, is another crucial condition to be verified.

Since photons interact very weakly among themselves, the only way they can thermalise is by coupling with matter. Therefore, the observed thermal equilibrium of the CMB can be explained if in the past there were processes able to couple radiation with matter. After *decoupling*, photons could then start to travel freely and, as we have seen, the expansion of the universe preserves their thermal equilibrium spectrum that we observe at present. In this way the CMB radiation can be regarded, within such *Hot Big Bang models*, as a *fossil radiation* left over from an early hot stage in the history of the universe.

12.2.2   Time-temperature relation (matter-dominated regime)

It is easy to derive how the temperature of the CMB evolves with time. In the matter-dominated regime, for $a_{\mathrm{eq}}^{M\Lambda} \gtrsim a \gtrsim a_{\mathrm{eq}} \simeq 2 \times 10^{-4}$ (see Eq. (10.39)) and for a flat universe, one has approximately

$$a(t) = a_{\mathrm{eq}}^{M\Lambda} \left(\frac{t}{t_{\mathrm{eq}}^{M\Lambda}} \right)^{\frac{2}{3}}, \qquad (12.6)$$

---

[1]This relation can be understood considering the meaning of the distribution function. This gives the mean occupation number of a quantum state with a certain momentum. Therefore, since the momentum of relic photons is changed only by the redshift due to the expansion, the number of photons at t in a quantum state with momentum $\vec{p}$ is equal to the number of photons today in a quantum state of momentum $\vec{p}_0 = a(t)\,\vec{p}$.

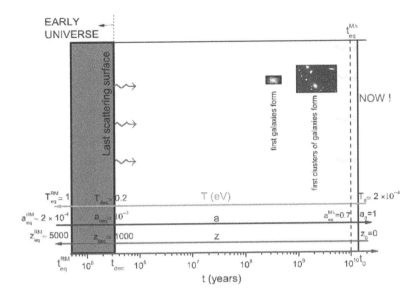

FIGURE 12.2: Schematic history of the universe from the matter-radiation equality time until present.

where $t_{eq}^{M\Lambda} \simeq 10.2\,\text{Gyr}$ is the matter-Λ equality time and $a_{eq}^{M\Lambda} \simeq 0.75$. Therefore, since $T(t) \propto a^{-1}(t)$, one has (for $t > t_{eq}$)

$$T(t) \simeq T(t_{eq}^{M\Lambda}) \left(\frac{t_{eq}^{M\Lambda}}{t}\right)^{\frac{2}{3}} \simeq 1.3 \times 10^3 \,\text{eV} \left(\frac{\text{yr}}{t}\right)^{\frac{2}{3}} \simeq 1.3 \times 10^8 \,\text{eV} \left(\frac{\text{s}}{t}\right)^{\frac{2}{3}} . \quad (12.7)$$

We have in this way a precise correspondence between cosmic time and temperature of CMB. Notice that the relation $T(t) \propto a(t)^{-1}$ is valid also before the decoupling between matter and radiation since in this case, as we already discussed, if thermal equilibrium holds, then $\varepsilon_R \propto T^4$, while on the other hand from the fluid equation, one has $\varepsilon_R \propto a^{-4}$.

A schematic description of the history of the universe after matter-radiation decoupling in terms of time and of temperature is shown in Fig. 12.2.

12.2.3 Baryon-to-photon ratio

Since relic photons were in thermal equilibrium throughout the history of the universe, at least starting from some initial time, their number density plays a special role, especially during the *early universe* stage. It provides a benchmark reference value for the number densities of all other particle species. In particular, an important cosmological physical quantity is the *baryon-to-photon ratio* defined as

$$\eta_B \equiv \frac{n_B}{n_\gamma} \simeq \frac{\Omega_B \, \varepsilon_c}{m_p \, c^2 \, n_\gamma}, \quad (12.8)$$

where, in the expression on the right-hand side, we used the approximation $m_n \simeq m_p$. This expression is valid below a certain temperature, when all baryonic number is transferred to nucleons.[2] This ratio can be evaluated at present and in this case, using respectively the numerical values Eqs. (6.42) and (2.34) for $\varepsilon_{c,0}$ and $n_{\gamma,0}$ and taking for the proton mass $m_p = 938.3\,\mathrm{MeV}/c^2$, one obtains

$$\eta_{B,0} \simeq 273.5\,\Omega_{B,0}h^2 \times 10^{-10}\,, \tag{12.9}$$

a useful relation since observations measure directly $\Omega_{B,0}h^2$ while $\eta_{B,0}$ is a physical quantity playing a direct role, as we will discuss, in Big Bang nucleosynthesis and baryogenesis models. As we will see, from the observations of CMB anisotropies one finds $\Omega_{B,0}h^2 \simeq 0.022$ corresponding to $\eta_{B,0} \simeq 6.1 \times 10^{-10}$.

Notice that if there is no process changing the number of photons and the baryonic number in a given comoving volume, i.e., $N_\gamma \equiv n_\gamma(t)\,a^3(t) = N_{\gamma,0}$ and $N_B \equiv n_B(t)\,a^3(t) = N_{B,0}$, then $\eta_B(t) = \eta_{B,0} = \mathrm{const}$. We will see in the next chapter that going sufficiently back in the history of the universe, one reaches such high temperatures that this result does not hold and η_B does change with time and, precisely, it was higher in the past. However, during the phase of the universe discussed in this chapter, we can safely assume $\eta_B(t) = \eta_{B,0} = \mathrm{const}$.

12.3 MATTER-RADIATION DECOUPLING AND RECOMBINATION

The observed CMB thermal spectrum can be elegantly explained by an early hot stage when radiation and matter where coupled together and the conditions for thermal equilibrium were satisfied. Thermal equilibrium spectrum would then remain imprinted until the present time. As we will see, this hot stage of matter-radiation coupling corresponds to values of the scale factor $a \lesssim 10^{-3}$, or, in terms of temperature, $T \gtrsim 0.1\,\mathrm{eV}$.

We have now to understand which physical processes could enforce matter-radiation coupling and when this stage came to an end, calculating the so-called matter-radiation decoupling time [8]. The value of the scale factor at matter-radiation decoupling places an upper limit on the initial value of the scale factor a_{in}. It should be clear that a_{in} can be very small and close to the singularity but it can never exactly vanish. In fact one expects that approaching the singularity, the validity of a cosmological description in terms of general relativity breaks down, since at the singularity the energy density diverges and the FRW metric becomes unphysical.

What process or processes could be responsible for the coupling of matter and radiation in the past? At present, ordinary matter is mainly constituted of neutral atoms. However, suppose that the Big Bang started with a value of the scale factor $a_{\mathrm{in}} \lesssim 10^{-5}$. From the redshift of momentum and the value of T_0, this corresponds to an average photon energy $\bar{E} \simeq 2.7\,T \gtrsim 30\,\mathrm{eV}$. The binding energy of hydrogen-1

---

[2]A derivation of the approximate expression is quite simple. For higher precision one should also take into account the relative abundance of neutrons compared to protons and the small difference between neutron and proton mass. However, this would produce just a small correction.

atoms, equivalent to its ionisation energy, is given by $Q \equiv (m_p + m_e - m_H) c^2 \simeq$ 13.6 eV, where $m_e \simeq 0.51$ MeV/c^2 is the electron mass and $m_H \simeq 938.8$ MeV/c^2 is the mass of the hydrogen-1 atom.[3] At these high energies it is then a good approximation to describe ordinary matter as a *plasma* of protons and electrons. As we will see in the next chapter, one should also consider the presence of helium nuclei but an account of their abundance introduces just a correction that we can safely neglect.

In this situation the main process, playing the role of a sort of glue between matter and radiation, is represented by *Thomson scattering*,

$$\gamma + e^- \leftrightarrow \gamma + e^- , \tag{12.10}$$

of photons on *free electrons*, barring the electrons bound in hydrogen atoms since in this case the scattering happens with hydrogen atoms as a whole (photoionisation) and the cross section is much smaller.

Thomson scattering exchanges and redistributes energy and momentum among photons and electron quantum states, a necessary condition for the thermalisation of the system.[4] The existence of such a hot stage can therefore potentially explain the thermal spectrum of CMB that we observe at present. One has to check that these processes are fast enough with respect to the expansion of the universe in order to be effective. This means that the mean free time between two subsequent scatterings has to be shorter than the age of the universe itself, otherwise scatterings are not frequent enough to bring photons into thermal equilibrium.

The photon mean free time τ_γ is trivially related to the photon mean free path λ_γ by $\tau_\gamma = \lambda_\gamma/c$. In the case of Thomson scattering one has

$$\lambda_\gamma = \frac{1}{n_e \, \sigma_T} \quad \Rightarrow \quad \tau_\gamma = \frac{1}{n_e \, \sigma_T \, c}, \tag{12.11}$$

where n_e is the number density of free electrons and σ_T is the Thomson scattering cross section given by

$$\sigma_T = \frac{8\pi}{3} \left(\frac{\alpha \, \hbar \, c}{m_e c^2} \right)^2 \simeq 6.65 \times 10^{-29} \, \text{m}^2 , \tag{12.12}$$

where $\alpha = e^2/(4\pi \, \hbar \, c) \simeq 1/137$ is the fine-structure constant in Lorentz–Heaviside units. As we discussed in detail, the age of the universe is approximately given by $t \sim H^{-1}$. Therefore, we have to impose

$$\tau_\gamma \lesssim t \sim H^{-1} \tag{12.13}$$

as a necessary condition for the collisions to be fast enough in the cosmological time

---

[3]The hydrogen-1 atom (symbol ^{1}H) is also called 'protium' and it is the most common of the hydrogen isotopes. The other two isotopes found in nature are hydrogen-2 (commonly called deuterium and indicated either with symbol ^{2}H or more usually with D) and hydrogen-3 (commonly called tritium and indicated with symbol ^{3}H or more usually with T).

[4]Electrons and protons couple with each other through fast electromagnetic interactions, so that Thomson scatterings effectively couple photons to all matter, not just to electrons.

scale.[5] In terms of the mean free path λ_γ, this condition is equivalent to writing

$$\lambda_\gamma \lesssim c\, H^{-1}\,, \tag{12.14}$$

that highlights another interesting way to understand it. On the right-hand side, we have the Hubble radius. Two comoving points at a larger distance will be pulled apart by the expansion at a speed higher than the speed of light. The condition Eq. (12.14) is therefore equivalent to saying that the mean free path has to be shorter than the Hubble radius otherwise, on average, a particle will never be able reach its target, since the expansion is dragging it away too fast.[6]

The condition (12.13) can also be simply recast in terms of the photon collision rate defined as $\Gamma_\gamma \equiv \tau_\gamma^{-1} = n_e\,\sigma_T\,c$, obtaining that this has to be higher than the expansion rate, explicitly

$$\Gamma_\gamma \gtrsim H\,. \tag{12.15}$$

On the other hand, when $\Gamma_\gamma \lesssim H$, the rate of Thomson scattering is too low and photons, on average, can be considered decoupled from matter. The *matter-radiation decoupling time* $t_{\rm dec}$ is defined as the critical time when $\Gamma(t_{\rm dec}) = H(t_{\rm dec})$.

Though we are discussing the coupling condition (12.15) in the specific case of Thomson scattering, it actually applies to a generic scattering process with rate Γ. If it is satisfied, i.e., $\Gamma \gtrsim H$, then the considered scattering process is said to be *on* (or *in equilibrium*), if it is not, i.e., $\Gamma \lesssim H$, then the scattering process is effectively *off* (or *out-of-equilibrium*) and it can be neglected for all practical purposes. The borderline case $\Gamma = H$ defines a decoupling time for the specific scattering process under consideration. Let us now go back to the case of Thomson scattering.

Let us first assume that all electrons and protons can be treated as free and that the density of hydrogen atoms can be neglected prior to the matter-radiation decoupling. In this case because of the electric charge neutrality of the universe, the number density of electrons is equal to the number density of protons and we can write

$$n_e = n_p \simeq n_B = \frac{n_{B,0}}{a^3} = \frac{\eta_{B,0}\,n_{\gamma,0}}{a^3} \simeq \frac{0.25\,{\rm m}^{-3}}{a^3}\,, \tag{12.16}$$

where in the last numerical expression we have used the numerical value in Eq. (2.34) for $n_{\gamma,0}$ and $\eta_{B,0} \simeq 6.1 \times 10^{-10}$. Inserting this expression for n_e and (12.12) for σ_e in Eq. (12.11), one obtains for the mean free time

$$\tau_\gamma = \tau_{\gamma,0}\,a^3\,, \quad \text{with } \tau_{\gamma,0} = (\eta_{B,0}\,n_{\gamma,0}\,\sigma_T\,c)^{-1} \simeq 0.2 \times 10^{21}\,{\rm s}\,. \tag{12.17}$$

We have now to impose the condition (12.13), comparing τ with the Hubble time

---

[5]Whether this is also a sufficient condition for equilibrium to hold, it has to be checked with kinetic equations able to track the evolution of particle distribution functions following continuously the transition between the coupled and the decoupled regime. It comes out that the criterion (12.13) represents a good instantaneous approximation providing a simple way to calculate the value of the temperature at the transition.

[6]Notice that this specific case nicely illustrates the general role played by the Hubble radius as instantaneous horizon, discussed in Section 10.1.

H^{-1}. Since $\Omega_0 \simeq 1$, the expansion parameter can be calculated from the Friedmann equation (6.49) neglecting the curvature term, obtaining

$$H^{-1}(a) = H_0^{-1} \left(\frac{\Omega_{R,0}}{a^4} + \frac{\Omega_{M,0}}{a^3} \right)^{-1/2} . \tag{12.18}$$

The condition for matter-radiation coupling (12.13) then gives easily

$$a^{\frac{3}{2}} \lesssim \frac{H_0^{-1}}{\tau_{\gamma,0}} \left(\frac{\Omega_{R,0}}{a} + \Omega_{M,0} \right)^{-\frac{1}{2}} . \tag{12.19}$$

Assuming that matter-radiation decoupling occurs in the matter-dominated regime, one can neglect the radiation contribution term, finding

$$a \lesssim a_{\rm dec} = \Omega_{M,0}^{-\frac{1}{3}} \left(\frac{H_0^{-1}}{\tau_{\gamma,0}} \right)^{\frac{2}{3}} \simeq 0.025 , \tag{12.20}$$

where we defined $a_{\rm dec} \equiv a(t_{\rm dec})$. This shows that matter-radiation decoupling does indeed occur in the matter-dominated regime since $a_{\rm eq} \simeq 2 \times 10^{-4}$ (see Eq. (10.39)). The condition on a can then be translated, from Eq. (12.5), into a condition on the temperature: $T \gtrsim 0.01\,{\rm eV}$.

This would be the matter-radiation decoupling temperature under the assumption that all electrons remain free for temperatures down to $T \sim 0.01\,{\rm eV}$. However, at such low temperatures, this assumption is clearly not valid.[7] Therefore, we need to revise our calculation, taking into account how the density of free electrons varies because of so-called *recombination*: while temperature drops during the expansion, electrons combine with protons to form hydrogen-1 atoms.[8]

Notice that the cross sections for the process of *photoionisation*

$$\gamma + {}^1{\rm H} \to p + e^- , \tag{12.21}$$

and for the inverse process, *radiative recombination*,

$$p + e^- \to \gamma + {}^1{\rm H} , \tag{12.22}$$

are much smaller than Thomson scattering cross section. This implies that while the free electron density decreases, the interaction rate between photons and matter also decreases and more and more photons start to propagate freely.

Therefore, we need to describe how the number density of free electrons changes taking into account recombination. This can be done introducing the *fractional ionisation X* defined as

$$X \equiv \frac{n_p}{n_p + n_{\rm H}} = \frac{n_e}{n_B} , \tag{12.23}$$

where $n_p = n_e$ is the number density of free protons and electrons and $n_{\rm H}$ is the

---

[7] Still this result is important, as we will notice in the next footnote.

[8] The name *recombination* can be misleading since within a Big Bang scenario electrons combine with protons for the first time!

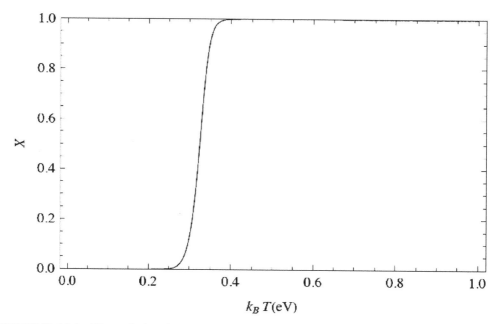

FIGURE 12.3: Plot of the *fractional ionisation* X versus temperature. At $T_{\text{rec}} \simeq 0.32\,\text{eV}$, one has $X(T_{\text{rec}}) = 0.5$.

number density of hydrogen-1 atoms. For $X = 0$, baryonic matter in the universe is entirely in the form of neutral atoms, while for $X = 1$ it is fully ionised.

If thermal equilibrium holds, then from statistical mechanics it can be shown that the variation of X while the temperature drops is described by a *Saha equation* given by (see Exercise 12.2)

$$\frac{1 - X}{X^2} \simeq 3.84\,\eta_{B,0} \left(\frac{T}{m_e c^2} \right)^{3/2} e^{Q/T} . \tag{12.24}$$

Notice that this equation is valid when electrons are non-relativistic, for $T \ll m_e c^2$. A plot of $X(T)$ is shown in Fig. 12.3. The *recombination temperature* T_{rec} is defined as $X(T_{\text{rec}}) = 1/2$. It marks the transition from a fully ionised plasma to a neutral gas made of hydrogen-1 atoms. Numerically, as one can see from Fig. 12.3, one finds

$$T_{\text{rec}} \simeq \frac{Q}{42} \simeq 0.32\,\text{eV} . \tag{12.25}$$

It should not surprise that $T_{\text{rec}} \ll Q$. This happens because of the very small value $\eta_{B,0} \sim 10^{-9}$, implying that there is a much greater number of photons than electrons. Therefore, even though the mean energy of photons is much lower than the ionisation energy Q, the number of photons at high energies, in the tail of the Planck distribution, is still sufficiently high to efficiently photoionise the hydrogen-1 atoms. From the solution of the Saha equation, we are now able to specify correctly the dependence $n_e(a)$ given by

$$n_e(a) = \frac{n_{B,0}}{a^3} X(a) . \tag{12.26}$$

It clearly drops faster than a^{-3}, since it takes into account both the expansion and recombination described by the function $X(a)$ that decreases while a increases. Imposing again the condition $\tau \lesssim H^{-1}$ this now leads to

$$a \lesssim a_{\text{dec}} = X(a_{\text{dec}})^{2/3} \, \Omega_{M,0}^{-\frac{1}{3}} \left(\frac{H_0^{-1}}{\tau_{\gamma,0}} \right)^{\frac{2}{3}} \simeq 0.025 \, X(a_{\text{dec}})^{2/3} \,, \qquad (12.27)$$

where we used $\Omega_{M,0} \simeq 0.31$ and the *Planck* result Eq. (2.26) for H_0. Using the solution for $X(a)$ coming from the Saha equation (12.24), one finds (see Exercise 12.3)[9]

$$a \lesssim a_{\text{dec}} \simeq 9.1 \times 10^{-4} \,, \quad z \gtrsim z_{\text{dec}} \simeq 1100 \,, \quad T \gtrsim T_{\text{dec}} \simeq 0.26 \, \text{eV} \,, \qquad (12.28)$$

respectively for the value of the scale factor, redshift and temperature at matter-radiation *decoupling*. Using the matter-dominated relation Eq. (12.6), one can also estimate the value of the age of the universe at decoupling, obtaining

$$t_{\text{dec}} \simeq \frac{t_{\text{eq}}^{M\Lambda}}{(a_{\text{eq}}^{M\Lambda} \, z_{\text{dec}})^{\frac{3}{2}}} \simeq 400,000 \, \text{yr} \,. \qquad (12.29)$$

Notice that the even after having accounted for recombination, decoupling still occurs after matter-radiation equality, in the matter-dominated regime. However, now the two events are much closer and for this reason, more precise results require to include also the radiation contribution. In particular the decoupling time has to be calculated solving the full integral Eq. (9.9). To this extent we need to calculate the precise value of $\Omega_{R,0}$ understanding the role of neutrinos. This will be discussed in detail in the next chapter, where we will obtain more precise values of all relevant quantities at decoupling.

On average t_{dec} is the time of the last scattering of photons with matter. However, it is clear that there is some dispersion around this value and the so-called *last scattering surface* is actually more a *last scattering layer*. One can expect that some photons are already free when $X = 1/2$ at $T \simeq T_{\text{rec}}$. A fuller treatment of recombination, would indeed show that the decoupling is not instantaneous and that recombination and decoupling take place gradually within the same period.

---

[9]The *Planck* collaboration [10] finds more precisely $z_{\text{dec}} \simeq 1090$ corresponding to $T_{\text{dec}} \simeq 0.256 \, \text{eV}$. They actually calculate the redshift $z_\star$ at the *time of last scattering* $t_\star$, defined as the time when the *optical depth* $\tau(t) \equiv \int_t^{t_0} \Gamma_\gamma(t') \, dt'$ is equal to unity. However, with very good approximation $z_{\text{dec}} \simeq z_\star$. There is an interesting twist in this story. Once recombination is over, all baryonic matter is in the form of neutral atoms. However, when stars and galaxies form, they emit high energy photons able to *reionise* atoms quite efficiently, so that today most of the atoms, mainly contained in intergalactic gas, are again ionised. Reionisation also occurs, like recombination, at a quite specific *reionisation time* t_{re} corresponding to a redshift z_{re}. Fortunately the optical depth remains much smaller than unity, otherwise we would not observe CMB anisotropies today since they would be smeared out by the scatterings of relic photons on ionised atoms. However, CMB anisotropies are sensitive to reionisation and the *Planck* collaboration finds $z_{\text{re}} = 8.8^{+1.7}_{-1.4}$. As noticed in the previous footnote, the redshift or reionisation corresponds to a temperature $T_{\text{re}} \simeq 0.002$ eV fortunately well below 0.01 eV, the temperature of decoupling of Thomson scatterings that we calculated assuming all electrons to be free. This is why $\tau(t_\star)$ remains much less than unity despite reionisation, otherwise we would not observe the CMB today.

The small difference between $T_{\text{rec}} \simeq 0.32\,\text{eV}$ and $T_{\text{dec}} \simeq 0.26\,\text{eV}$ gives an idea of the interval of temperatures corresponding to the recombination-decoupling stage.

We can then conclude saying that the Planckian spectrum of the CMB can be explained by a Big Bang model with an initial value of the scale factor $a_{\text{in}} \lesssim 10^{-3}$ corresponding to an initial temperature $T_{\text{in}} \gtrsim 0.3\,\text{eV}$.

However, this condition is not sufficient. It can be indeed shown that one has also to require a condition $\eta_B^{\text{dec}} \equiv \eta_B(t_{\text{dec}}) \ll 1$, i.e., that the number of baryons per number of photons has to be a very small number, something nicely confirmed by the observations finding, as we have seen, $\eta_B^{\text{dec}} = \eta_{B,0} \sim 6 \times 10^{-10}$: theory and experiments are in this way perfectly consistent with each other!

In conclusion, the CMB thermal equilibrium is a strong indication that an early hot stage, when matter and radiation were coupled, occurred in the past. The value of the scale factor had to be necessarily smaller than $a_{\text{dec}} \sim 10^{-3}$, relatively close to the singularity, and this justifies the name of *Hot Big Bang model*.

12.4 TEMPERATURE ANISOTROPIES

Let us now discuss the main features of the observed CMB temperature anisotropies and how in recent years, with the discovery of the *acoustic peaks* in the power spectrum of primary anisotropies, it became possible to extract precise values of the cosmological parameters. For a given direction of sky $\hat{n}$ described by polar coordinates (θ, ϕ), the *temperature fluctuation* is defined as

$$\frac{\Delta T}{T}(\hat{n}) \equiv \frac{T(\hat{n}) - \langle T \rangle}{\langle T \rangle}. \tag{12.30}$$

One can also introduce the *temperature correlation function* defined as[10]

$$C(\Delta\theta) = \left\langle \frac{\Delta T}{T}(\hat{n}) \frac{\Delta T}{T}(\hat{n}') \right\rangle_{\hat{n}\cdot\hat{n}'=\cos\Delta\theta}, \tag{12.31}$$

where $\Delta\theta$ is the angular separation between two directions $\hat{n}$ and $\hat{n}'$. Experiments have a finite angular resolution and, therefore, they can measure the correlation function only on angular scales larger than the resolution of the instrument.

Temperature fluctuations can be expanded in *spherical harmonic functions* $Y_{lm}(\theta, \phi)$,

$$\frac{\Delta T}{T}(\hat{n}) = \sum_{\ell=0}^{\infty} \sum_{m=-\ell}^{m=\ell} a_{\ell m}\, Y_{\ell m}(\theta, \phi). \tag{12.32}$$

Spherical harmonics are related to the associated Legendre polynomials by

$$Y_{\ell m}(\theta, \phi) = \sqrt{(2\ell + 1)\frac{(\ell - m)!}{(\ell + m)!}}\, P_{\ell m}(\cos\theta)\, e^{i\,m\,\phi}. \tag{12.33}$$

---

[10]The average here has to be meant over all possible choices $\hat{n}, \hat{n}'$. This is a two-point correlation function. One can also introduce a three-point or higher order correlation function. However, for the typical anisotropies predicted by inflation, a random Gaussian homogeneous and isotropic field, these vanish within cosmic variance uncertainty.

Since we subtracted the average value of the temperature, the term $\ell = 0$, corresponding to the average of $\Delta T/T$ over all sky, vanishes. The $l = 1$ moment corresponds to the dipole term that, as discussed, is generated by the Doppler shift due to the motion of the Earth with respect to the comoving system. Moments with $\ell \geq 2$ are cosmologically the most interesting ones, since they correspond to the *primary anisotropies* that have their origin in the early universe [9].

The cosmological principle implies that there are no privileged directions in the universe and therefore temperature fluctuations have to be statistically isotropic implying that the correlation function can depend only on their angular size $\Delta\theta$ but not on their orientation $\Delta\phi$. This implies that if we insert the expansion Eq. (12.32) into the definition of correlation function, the dependence on m has to cancel out[11] so that the correlation function can be expressed in terms of Legendre polynomials (the associated Legendre polynomials with $m = 0$),

$$C(\Delta\theta) = \sum_{\ell} \frac{2\ell + 1}{4\pi} C_\ell \, P_\ell(\cos\Delta\theta) \,. \tag{12.34}$$

The C_ℓ's are the *multipole moments* and their set is the so-called *angular power spectrum*.[12] The multipole moments are an average over m of $\langle |a_{\ell m}|^2 \rangle$, explicitly

$$C_\ell = \frac{1}{2\ell + 1} \sum_{m=-\ell}^{\ell} \langle |a_{\ell m}|^2 \rangle \,. \tag{12.35}$$

Each multipole is a measure of the temperature fluctuation on an angular scale

$$\Delta\theta = \frac{180°}{\ell} \,. \tag{12.36}$$

CMB observations are customarily presented plotting the quantity

$$\Delta_T \equiv \langle T \rangle \sqrt{\frac{\ell(\ell + 1)}{2\pi} C_\ell} \tag{12.37}$$

---

[11]This can be seen considering that, more specifically, isotropy implies $\langle a_{\ell m} \, a_{\ell',m'}^* \rangle = \delta_{\ell\ell'} \, \delta_{mm'} \, C_\ell$ and from this condition one can derive (12.34) and (12.35).

[12]More precisely this is the temperature-temperature angular power spectrum, often simply referred to as the TT spectrum. CMB experiments such as WMAP and *Planck* also observed polarisation anisotropies. These have two different modes called E-mode and B-mode. One can then also measure, in addition to TT spectrum, the TE, EE and BB *cross power spectra*. The other two (EB, TB) vanish due to simple parity considerations. Here we will focus on the TT power spectrum. A more detailed discussion including also polarisation power spectra can be found in [11]. We just mention that while E-mode polarisation and its anisotropies have been well measured, B-mode polarisation, of cosmological origin, has so far escaped detection. This can be produced only by tensor primordial perturbations, that would be the imprint on CMB of primordial gravitational waves. Inflationary models predict some positive signal but its size is strongly model dependent. Observations have so far placed an upper bound on the tensor-to-scalar ratio of primordial perturbations (see discussion at page 207). This translates into important constraints on models of inflation.

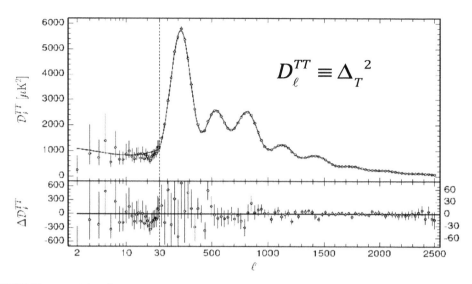

FIGURE 12.4: The CMB acoustic peaks in temperature anisotropies from the latest *Planck* 2015 data (upper panel). The solid line is the best fit within the ΛCDM model (that is a flat model, i.e., $\Omega_0 = 1$). In the lower panel the plot shows the residuals with respect to the best fit. It can be noticed how there are some appreciable discrepancies for $\ell \simeq 20$, but the global fit is nevertheless very good.

or its square. The WMAP satellite, with a resolution $\Delta\theta_{\min} \simeq 10'$, could measure the C_ℓ's up to $\ell_{\max} \simeq 1300$, finding that the C_ℓ's (the angular power spectrum) exhibit a series of peaks characterised by their position at some value of ℓ and by their height [5]. The *Planck* satellite [7] has more recently been able to measure, thanks to a much higher resolution $\Delta\theta \simeq 3'$, the angular power spectrum up to $\ell_{\max} \simeq 2500$.

In Fig. 12.4, we show the results from the latest (2015) *Planck* data release [10]. One can see that the quantity Δ_T^2 exhibits a series of *acoustic peaks* characterised by a position at a given ℓ and by a certain height. One can also notice a solid line providing a very good fit to the data. The solid line is the best fit derived within the ΛCDM model, corresponding to best fit values for the parameters of the model.

In the next subsection, we will give a qualitative discussion on the physics of the acoustic peaks[13] and show how the position of the first peak allows a determination of the curvature of the universe. More generally, the positions and heights of the acoustic peaks depend on different cosmological parameters. The ΛCDM model, also conventionally called the *baseline model*, is a minimal model depending on just six cosmological parameters. It is a flat Lemaitre model ($\Omega_0 = 1$) and two of the six primary parameters used to fit the power spectrum are $\Omega_{B,0}\, h^2$ and $\Omega_{DM,0}\, h^2$. The other four primary parameters are: i) The optical depth $\tau(t) \equiv \int_t^{t_0} \Gamma_\gamma(t')\, dt'$, ii) The angular size of the sound-horizon at decoupling $\theta_{\mathrm{dec}}^{\mathrm{sh}}$ that we will introduce

---

[13]Acoustic peaks in CMB temperature anisotropies were predicted theoretically much earlier than their experimental discovery [6].

TABLE 12.1 Cosmological parameters from *Planck* 2015 data [10].

Parameter	TT+lowP	TT+lowP+lensing	TT,TE,EE+low P+lensing+ext
$\Omega_{B,0}\,h^2$	0.02222 ± 0.00023	0.02226 ± 0.00023	0.02230 ± 0.00014
$\Omega_{DM,0}\,h^2$	0.1197 ± 0.0022	0.1186 ± 0.0020	0.1188 ± 0.0010
$100\,\theta_{\text{dec}}^{\text{sh}}$	1.04085 ± 0.00047	1.04103 ± 0.00046	1.04093 ± 0.00030
τ	0.078 ± 0.019	0.066 ± 0.016	0.066 ± 0.012
$\ln(10^{10}\,A_s)$	3.089 ± 0.036	3.062 ± 0.029	3.064 ± 0.023
n_s	0.9655 ± 0.0062	0.9677 ± 0.0060	0.9667 ± 0.0040
h	0.6731 ± 0.0096	0.6781 ± 0.0092	0.6774 ± 0.0046
$\Omega_{\Lambda,0}$	0.685 ± 0.013	0.692 ± 0.012	0.6911 ± 0.0062
$\Omega_{M,0}$	0.315 ± 0.013	0.308 ± 0.012	0.3089 ± 0.0062
$\Omega_{M,0}\,h^2$	0.1426 ± 0.0020	0.1415 ± 0.0019	0.14170 ± 0.00097
t_0/Gyr	13.813 ± 0.038	13.799 ± 0.038	13.807 ± 0.021
z_{re}	$9.9^{+1.8}_{-1.6}$	$8.8^{+1.7}_{-1.4}$	$8.8^{+1.2}_{-1.1}$
z_{dec}	1090.09 ± 0.42	1089.94 ± 0.42	1089.90 ± 0.23
z_{eq}	3393 ± 49	3365 ± 44	3371 ± 23

Note: The first six parameters are the primary parameters used in the fit within the ΛCDM model (the baseline model). The other parameters are derived parameters. In the first column only information from TT spectrum and low ℓ polarisation data is used. In the second column information from gravitational lensing is also added. Finally, in the last column also polarisation data at high multipoles and external (HST H_0+ JLA SNIa +BAO) data is added.

in the next subsection; iii) The amplitude A_s and iv) the *spectral index n_s* of the scalar primordial power spectrum. The scalar primordial power spectrum is the result of those initial perturbations that are necessary in order to understand all inhomogeneities we observe, including the CMB temperature anisotropies. Within the ΛCDM model, these are generated by quantum fluctuations during inflation.

From the primary parameters one can derive all other cosmological parameters, including: Hubble constant H_0, age of the universe t_0, $\Omega_{\Lambda,0}$, the value of the redshift at matter-radiation equality time z_{eq} and the redshift at reionization z_{re}.

Their best fit values with 68% errors derived from the *Planck* satellite 2015 data are shown in Table 12.1 for different choices of the datasets as explained in the table note.[14] The acoustic peaks are particularly sensitive to parameters entering the properties of the universe at recombination. This is because CMB is a kind of snapshot of the universe at the time of recombination. In particular, they are particularly sensitive to the value of the baryon abundance $\Omega_{B0}\,h^2$, determining the speed of the acoustic waves in the primordial plasma at recombination. As one

---

[14]We used as reference values in the book those from the second column, obtained combining TT power spectrum, polarisation at low multipoles ('lowP') and information from gravitational lensing (this is the gravitational lensing experienced by the photons crossing the universe from the last scattering surface to us and gives information on the distribution of matter in the universe).

can see from the second column in the table, the *Planck* collaboration finds an extremely precise value given by

$$\Omega_{B,0}^{(CMB)} h^2 = 0.02222 \pm 0.00023 \,. \tag{12.38}$$

When this is combined with the value of h (see Eq. (2.25)), one finds

$$\Omega_{B,0}^{(CMB)} = 0.0490 \pm 0.0012 \,, \tag{12.39}$$

showing that ordinary matter contributes approximately only 5% to the total energy density of the universe. Moreover, using the relation (12.9), one finds for the baryon-to-photon ratio

$$\eta_{B,0}^{(CMB)} = (6.08 \pm 0.06) \times 10^{-10} \,. \tag{12.40}$$

Another quantity that is determined relatively well, is the total amount of matter,

$$\Omega_{M,0}^{(CMB)} = 0.308 \pm 0.012 \,, \tag{12.41}$$

confirming again the value deduced from SN type Ia data and dynamics of clusters of galaxies. The results from CMB anisotropies confirm in a robust and direct way the indirect evidence discussed at the end of Chapter 11: there is a dark matter component of non-baryonic origin, or some gravitational effect able to mimic it, that points to the existence of a new fluid, unexplained within the SM, or to a modification of gravity beyond general relativity. In Chapter 17, we will discuss in detail possible explanations for this mysterious dark matter component, one of the most fascinating mysteries in modern science.

Notice that when the information from temperature anisotropies is combined with additional information, the errors reduce considerably in the case of different quantities. First of all, both the WMAP and *Planck* satellites have been able to measure, in addition to temperature anisotropies, also the anisotropies of CMB polarization.[15] Very recently, the *Planck* satellite has also been able to measure the gravitational lensing effects induced by structures along the path of CMB photons from the last scattering surface to us. In this way CMB is able to provide information on the properties of the universe beyond just the recombination time.

Another cosmological observation that is also useful to reconstruct the history of the universe expansion is represented by so-called *baryon acoustic oscillations* (BAO), analogous to CMB acoustic peaks in galaxy power spectra. This is measured in galaxy surveys and the information can be combined with that from CMB (see last column in Table 12.1).

One can also add additional parameters to the minimal set of the ΛCDM model, checking whether a better fit of the data is obtained. The result is that so far the ΛCDM model provides the best fit when the information from CMB anisotropies is combined with external information such as BAO, type Ia supernovae and a prior $H_0 = (70.6 \pm 3.3)$ km s^{-1} Mpc^{-1} from astrophysical measurements ('ext' in the last

---

[15]It can be shown indeed that CMB photons have a non zero polarization.

TABLE 12.2 95%C.L. constraints on 1-parameter extensions to the ΛCDM model from *Planck* 2015 data [10].

Parameter	TT+lowP	TT+lowP+lensing	TT,TE,EE+low P+lensing+ext
$\Omega_{k,0} \equiv 1 - \Omega_0$	$-0.052^{+0.049}_{-0.055}$	$-0.005^{+0.016}_{-0.017}$	$0.0008^{+0.0040}_{-0.0039}$
$\sum_i m_{\nu_i}/\mathrm{eV}$	< 0.715	< 0.675	< 0.194
N_ν^{dec}	$3.13^{+0.64}_{-0.63}$	$3.13^{+0.62}_{-0.61}$	3.04 ± 0.33
Y_p	$0.252^{+0.041}_{-0.042}$	$0.251^{+0.040}_{-0.039}$	$0.249^{+0.025}_{-0.026}$
$dn_s/d\ln k$	-0.008 ± 0.016	-0.003 ± 0.015	-0.002 ± 0.013
r	< 0.103	< 0.114	< 0.113
w	$-1.54^{+0.62}_{-0.50}$	$-1.41^{+0.64}_{-0.56}$	$-1.019^{+0.075}_{-0.080}$

column of Tables 12.1 and 12.2). In this way one derives stringent constraints on the additional considered parameters. In Table 12.2, we show the 95% C.L. constraints on different interesting quantities for the same datasets of Table 12.1 in the case of 1-parameter extensions to the CDM model. It should be noticed how this time the additional external information drastically reduces the errors on some of the parameters, in particular on $\Omega_{k,0}$, w and the upper bound on the sum of neutrino masses $\sum_i m_{\nu_i}$. One can see how the deviation from flatness $|\Omega_{k,0}|$ is constrained to be less than 1%. In this case one can see how gravitational lensing reduces the error by ten times. Very interestingly, one can also consider neutrino masses. One can see from the table how the most recent data allow placing a stringent upper bound on the sum of the neutrino masses $\sum_i m_{\nu_i} < 0.194\,\mathrm{eV}$ at 95%C.L., when external information is combined (last column), this is the most stringent experimental constraint on the so-called *absolute neutrino mass*.[16]

12.4.1 Position of the first acoustic peak

A precise derivation of the acoustic peaks involves quite advanced tools of relativistic fluid-dynamics and goes beyond the scope of this book.[17] However, there is a very important feature that can be understood in a simple way: the position of the first peak and how this depends on the value of Ω_0 and, therefore, how its position determines the curvature of the universe.

In order to explain the temperature fluctuations one has to assume, going beyond a strict application of the cosmological principle, the existence of an initial spectrum of *primordial perturbations*. At this stage we can consider this as an assumption on the initial conditions. Primordial perturbations are mainly carried by

---

[16]On the other hand consider that neutrino mixing experiments place a lower bound $\sum_i m_{\nu_i} \gtrsim 0.06\,\mathrm{eV}$ at 95%C.L. In the next years, new cosmological data should improve the sensitivity to neutrino masses down to $\delta(\sum_i m_{\nu_i}) \sim 0.02\,\mathrm{eV}$, meaning that either they will be able to detect a signal compatible with neutrino mixing experiments, or that the ΛCDM model should be modified in some way in order to remove the clash with neutrino mixing experiments. In both cases, this represents a fascinating test of the compatibility between cosmology and neutrino physics.

[17]For a comprehensive analytical discussion on the physics of the acoustic peaks see [11]. For a numerical derivation typically codes like CMBFAST [12] and its updated version CAMB [13] are adopted.

dark matter since, by definition, their mass is large enough that they can be considered basically at rest in the comoving volume. Propagating particles like photons tend to smooth the perturbations by *free streaming*. This of course shows why dark matter plays a crucial role in a correct description of the evolution of perturbations and at the same time how dark matter should be present already in the early universe, certainly during recombination.

The existence of perturbations generates gravitational attraction and instability in ordinary matter and in radiation as well (in general relativity light also feels the gravitational attraction). Before and during recombination we have therefore a fluid of coupled photons, electrons and protons.[18] Under the simultaneous influence of the attractive gravitational force from potential wells contrasted by pressure, the fluid undergoes cyclic compression and expansion phases, so-called acoustic oscillations. These generate sound waves that propagate in the fluid dominated by radiation.[19] Sound waves propagate with a *sound speed* given by

$$c_{\text{s}} = \sqrt{w_{\text{pb}}}\, c \simeq \frac{c}{\sqrt{3}}, \qquad (12.42)$$

where w_{pb} gives the equation of state for a photon-baryon fluid, $p = w_{\text{pb}}\, \varepsilon$, and, because of the much higher number of photons, one has $w_{\text{pb}} \simeq 1/3$, the usual value holding for radiation.[20]

The maximum wave length of the acoustic oscillations is set by the size of a causally connected region at decoupling. Since causal connection is in this case due to sound propagation, the size will be given by the so-called *sound horizon*, d_{sh}, at decoupling. This can be obtained from the horizon at present (see Eq. (10.4)) simply replacing the speed of light c with the sound speed $c_{\text{s}} = c/\sqrt{3}$ and t_0 with t_{dec}.

It is convenient to express the sound horizon at decoupling in comoving length, equivalent to rescale it at the present taking into account the expansion. As an analogy, one can think of the sound horizon as a cosmic soundbox and the acoustic peaks correspond to its fundamental harmonics. We can then write

$$d_{\text{sh}}^{(0)}(t_{\text{dec}}) = c_s \int_0^{t_{\text{dec}}} \frac{dt}{a} \;\Rightarrow\; d_{\text{sh}}^{(0)}(a_{\text{dec}}) = c_s \int_0^{a_{\text{dec}}} \frac{da}{\dot{a}\, a} = c_s \int_0^{a_{\text{dec}}} \frac{da}{H(a)\, a^2}.$$
$$(12.43)$$

The integral on the right-hand side can be decomposed in the sum of a contribution from the radiation-dominated regime plus a contribution from the matter-dominated regime. From the values of a_{dec} and a_{eq}, one finds that the latter is about three times larger and, therefore, one can approximate the whole integral in

---

[18]Notice that protons couple to electrons through electromagnetic interactions and, therefore, they are also indirectly coupled to photons.

[19]Notice that in this case one can literally talk of the *sound of light*!

[20]There is however a small correction from baryons that introduces a dependence on $\Omega_{B,0}h^2$: it is this dependence that translates into a sensitivity of the acoustic peak to $\Omega_{B,0}h^2$ and into the precise determination of $\Omega_{B,0}h^2$ Eq. (12.38). Notice also that the contribution to w_{pb} from electrons is negligible compared to protons because they are much lighter, this is why it is possible to talk of a photon-baryon fluid.

the matter-dominated regime. From the Friedmann equation, written in the form (12.18), one has $H \simeq H_0 \sqrt{\Omega_{M,0}}/a^{3/2}$, finding

$$d_{\rm sh}^{(0)}(a_{\rm dec}) \simeq \frac{c_s H_0^{-1}}{\sqrt{\Omega_{M,0}}} \int_0^{a_{\rm dec}} \frac{da}{a^{1/2}} = 2 \frac{c_s H_0^{-1}}{\sqrt{\Omega_{M,0}}} \sqrt{a_{\rm dec}} \,. \tag{12.44}$$

The first peak corresponds to the angular size of the sound horizon at the time of matter-radiation decoupling and this is simply given by the ratio of the comoving sound horizon distance at $t_{\rm dec}$ to the distance of the last scattering surface at the present time. The distance of the last scattering surface is basically given by the horizon distance at present $d_{H,0}$ in a Lemaitre model. Assuming a *flat universe* and neglecting the Λ-dominated period, one finds that the sound horizon at decoupling subtends an angle

$$\theta_{\rm dec}^{\rm sh} = \frac{d_{\rm sh}^{(0)}(a_{\rm dec})}{d_{H,0}} \simeq \frac{c_s}{c} \sqrt{a_{\rm dec}} \,. \tag{12.45}$$

The first peak will then correspond to

$$\ell_{\rm 1st\,peak} = \frac{\pi}{\sqrt{a_{\rm dec}}} \frac{c}{c_s} \simeq \pi \sqrt{3\, z_{\rm dec}} \,. \tag{12.46}$$

Considering that $z_{\rm dec} \simeq 1100$, one finds

$$\ell_{\rm 1st\,peak} \simeq 180 \,, \tag{12.47}$$

that is in good agreement with the position of the first peak, $\ell_{\rm 1st\,peak}^{\rm exp} \simeq 220$, found by the *Planck* satellite and shown in Fig. 12.4.[21] In a curved space, the distance to the horizon has to be more generally replaced by the *angular distance to the horizon* and the expression for the position of the first peak is modified becoming (see Exercise 12.5)

$$\ell_{\rm 1st\,peak}(\Omega_0) \simeq \frac{\ell_1(\Omega_0 = 1)}{\sqrt{\Omega_0}} \,. \tag{12.48}$$

Therefore, a measurement of $\ell_{\rm 1st\,peak}$ translates straightforwardly into a measurement of Ω_0. Since the observations have found $\ell_{\rm 1st\,peak} \simeq \ell_1(\Omega_0 = 1)$ [14], they imply that the universe is very close to being flat, as one can see from the result shown in Table 12.2 from 2015 *Planck* data.

EXERCISES

Exercise 12.1 *What is the mean free time of a photon when $T \simeq 1\,{\rm eV}$?*

Exercise 12.2 *Derive the Saha equation 12.24 describing the evolution of the fractional ionisation with temperature.*

---

[21] A more precise calculation of $\theta_{\rm dec}^{\rm sh}$ has to take into account the contribution from the radiation-dominated regime in $d_{\rm sh}^{(0)}(a_{\rm dec})$ and from the Λ-dominated regime in $d_{H,0}$ (see Exercise 12.4). At the same time one has to take into account a more subtle effect called *radiation driving* that causes a shift of ℓ_1 compared to $\pi/\theta_{\rm dec}^{\rm sh}$ [11].

Exercise 12.3 *Derive the value of redshift at decoupling as in Eq. (12.28).*

Exercise 12.4 *Derive a more precise expression of the angular size of the sound horizon at decoupling taking into account radiation and Λ-dominated regimes.*

Exercise 12.5 *Derive the dependence of ℓ_1 on Ω_0 as in Eq. (12.48).*

BIBLIOGRAPHY

[1] J. C. Mather et al., *Measurement of the cosmic microwave background spectrum by the COBE FIRAS instrument*, Astrophys. J. **420** (1994) 439.

[2] D. J. Fixsen, *The temperature of the cosmic microwave background*, Astrophys. J. **707** (2009) 511 [arXiv:0911.1955].

[3] Conklin, E. K., *Velocity of the Earth with respect to the cosmic background radiation*, Nature, **222** (1969) 971.

[4] C. L. Bennett et al., *Four year COBE DMR cosmic microwave background observations: Maps and basic results*, Astrophys. J. **464** (1996) L1.

[5] C. L. Bennett et al. [WMAP Collaboration], *Nine-year Wilkinson microwave anisotropy probe (WMAP) observations: Final maps and results*, Astrophys. J. Suppl. **208** (2013) 20 [arXiv:1212.5225 [astro-ph.CO]].

[6] A. D. Sakharov, Zh. Eksp. Teor. Fiz. **49** no.1, 345 [Sov. Phys. JETP **22** (1966) 241]; J. Silk, *Cosmic black body radiation and galaxy formation*, Astrophys. J. **151** (1968) 459; P. J. E. Peebles and J. T. Yu, *Primeval adiabatic perturbation in an expanding universe*, Astrophys. J. **162** (1970) 815; R. A. Sunyaev and Y. B. Zeldovich, *Small scale fluctuations of relic radiation*, Astrophys. Space Sci. **7** (1970) 3.

[7] P. A. R. Ade et al. [Planck Collaboration], *Planck 2013 results. XVI. Cosmological parameters*, Astron. Astrophys. **571** (2014) A16 [arXiv:1303.5076 [astro-ph.CO]].

[8] P. J. E. Peebles, *Recombination of the primeval plasma*, Astrophys. J. **153** (1968) 1; Y. B. Zeldovich, V. G. Kurt and R. A. Sunyaev, *Recombination of hydrogen in the hot model of the universe*, Sov. Phys. JETP **28** (1969) 146 [Zh. Eksp. Teor. Fiz. **55** (1968) 278].

[9] For a detailed discussion see S. Weinberg, *Cosmology*, Oxford, UK: Oxford Univ. Pr. (2008) 593 p.

[10] P. A. R. Ade et al. [Planck Collaboration], *Planck 2015 results. XIII. Cosmological parameters*, Astron. Astrophys. **594** (2016) A13 [arXiv:1502.01589].

[11] W. Hu and S. Dodelson, *Cosmic microwave background anisotropies*, Ann. Rev. Astron. Astrophys. **40** (2002) 171 doi:10.1146/annurev.astro.40.060401.093926 [astro-ph/0110414].

[12] U. Seljak and M. Zaldarriaga, *A Line of sight integration approach to cosmic microwave background anisotropies*, Astrophys. J. **469** (1996) 437 [astro-ph/9603033].

[13] A. Lewis, *Efficient sampling of fast and slow cosmological parameters*, Phys. Rev. D **87** (2013) no.10, 103529 [arXiv:1304.4473 [astro-ph.CO]].

[14] P. de Bernardis et al. [Boomerang Collaboration], *A flat universe from high resolution maps of the cosmic microwave background radiation*, Nature **404** (2000) 955 [astro-ph/0004404].

CHAPTER 13

Radiation-dominated regime

In this chapter we continue our journey backwards in time, toward hotter and hotter phases in the history of the universe. Elementary particle physics plays a crucial role in the early universe during the radiation-dominated regime. In the first section we introduce the number of ultra-relativistic degrees of freedom, determining the expansion rate at a given temperature. This quantity provides an important (and historical) connection between cosmological observables and particle physics. Analogously to the case of matter-radiation decoupling, we discuss how neutrinos decouple from the thermal bath at a temperature $T_\nu^{\rm dec} \sim 1\,{\rm MeV}$. The CMB acoustic peaks and, as we will discuss in the next chapter, primordial nuclear abundances provide strong evidence that three neutrino species were indeed present during Big Bang nucleosynthesis and recombination. This implies that the initial temperature of the radiation-dominated regime had to be at least higher than $T_\nu^{\rm dec}$. Though there is no compelling experimental evidence supporting higher values, there are strong reasons to believe that this had to be even higher than the electroweak scale $E_{EW} \sim 100\,{\rm GeV}$. At these high temperatures, all standard model particle species would thermalise, and for this reason we calculate the number of ultra-relativistic degrees of freedom up to these very high temperatures: the early universe can indeed be regarded as a very special high energy collider!

13.1 NUMBER OF ULTRA-RELATIVISTIC DEGREES OF FREEDOM

As we have seen, the matter-radiation equality time corresponds to a value of the scale factor $a_{\rm eq} \sim 10^{-4}$ and temperature $T_{\rm eq} \sim 1\,{\rm eV}$. Notice that the matter-radiation decoupling occurs at a slightly lower temperature $T_{\rm dec} \simeq 0.26\,{\rm eV}$. The history of the early universe is then almost entirely occurring in the radiation-dominated regime on a logarithmic scale.

In this regime the relation between time and temperature can be easily found directly from the Friedmann equation, since the curvature term can be safely neglected, obtaining

$$H^2 \simeq \frac{8\pi\,G}{3\,c^2}\,\varepsilon_R\,, \tag{13.1}$$

where we neglected the matter component. The energy density of radiation can be written, in terms of temperature, as

$$\varepsilon_R(T) = g_R(T) \frac{\pi^2}{30} \frac{T^4}{(\hbar c)^3},\tag{13.2}$$

where $g_R(T)$ is the *number of ultra-relativistic degrees of freedom* at temperature T. Therefore, the expansion rate H at temperature T can be expressed as

$$H(T) = \sqrt{\frac{8\pi^3 g_R(T)}{90}} \frac{\hbar c}{M_P c} \left(\frac{T}{\hbar c}\right)^2 \simeq 0.21 \sqrt{g_R(T)} \left(\frac{T}{\text{MeV}}\right)^2 s^{-1},\tag{13.3}$$

where we replaced G with the Planck mass using its definition Eq. (2.10). Considering that in the radiation-dominated regime $a(t) \propto t^{1/2}$, one has straightforwardly

$$H(t) = \frac{\dot{a}}{a} = \frac{d\ln a}{dt} = \frac{1}{2t}.\tag{13.4}$$

Inserting this equation and Eq. (13.2) into the Friedmann equation, we find the time-temperature relation in the radiation-dominated regime

$$t = \frac{1}{2} \sqrt{\frac{90}{8\pi^3 g_R(T)}} \frac{M_P c}{\hbar c} \left(\frac{\hbar c}{T}\right)^2 \simeq \frac{2.4\,\text{s}}{\sqrt{g_R(T)}} \left(\frac{\text{MeV}}{T}\right)^2.\tag{13.5}$$

If the only ultra-relativistic particles were photons, then simply $g_R = 2$. However, the early universe behaves as a sort of particle collider but with an important difference. In a particle collider the beam energy is fixed by the experimental setup. For example, in the LHC two collinear anti-circulating proton beams with equal energy collide. The planned energy per beam is $7\,\text{TeV}$, implying a centre mass energy of $14\,\text{TeV}$. This means that, in the collisions, new particles with a mass not higher than $14\,\text{TeV}$ can be produced.

In the early universe the energy of a generic particle species X is not fixed, but has to be described by its distribution function $f_X(\vec{p}, t)$, defined as the mean occupation number of a quantum state with momentum $\vec{p}$. All calculations have to be done averaging on the distribution function. If the particle species X is in thermal equilibrium, then *ultra-relativistic* thermal equilibrium applies if $T \gg m_X c^2/2$, where m_X is the mass of X.[1] Moreover, since in this case $p_X \simeq \varepsilon_X/3$, the X particles contribute to radiation.[2] More precisely, let us indicate with X_b a generic bosonic

---

[1]Notice that the condition $T \gg m_X c^2/2$ is not sufficient for thermal equilibrium to hold, it just implies that if the X's are in thermal equilibrium, then they can be treated as ultra-relativistic. For thermal equilibrium it is necessary that the total interaction rate of the X's is in equilibrium. This, as we will see, will play an important role for neutrinos. Notice also that as a central value for the temperature for the transition from non-relativistic to ultra-relativistic equilibrium, it is more precise to use half the mass of the particle rather than just the mass. At $T \simeq m_X c^2$ an ultra-relativistic approximation is already quite good (see Exercise 13.4).

[2]On the other hand, in the limit $T/(m_X c^2) \to 0$, a particle species becomes pressure-less and contributes to matter if there is a non-vanishing asymmetry between particles and anti-particles, as in the case of electrons and positrons that we are just going to discuss. If it has a vanishing asymmetry, then one has to calculate the so-called relic abundance that will contribute to matter. This point will be particularly relevant when we discuss dark matter in Chapter 18.

particle species (such as photons) and with X_f a generic fermionic particle species (such as electrons and positrons). In this case the contribution to the radiation component of the energy density is given respectively by

$$\varepsilon_{X_b} = g_{X_b} \frac{\pi^2}{30} \frac{T^4}{(\hbar c)^3} \quad \text{and} \quad \varepsilon_{X_f} = \frac{7}{8} g_{X_f} \frac{\pi^2}{30} \frac{T^4}{(\hbar c)^3}, \tag{13.6}$$

where $g_X = (2S + 1)$ and S is the spin of the particle.[3] The different result for bosons and fermions, encoded in the factor $7/8$, is simply a consequence of the different sign in the denominator of equilibrium distribution functions: a minus sign in the Bose–Einstein distribution function applying to bosons, a plus sign in the Fermi–Dirac distribution function applying to fermions. In this case, for all particles in ultra-relativistic thermal equilibrium, implying $T \gg m_{X_{b,f}} c^2/2$, one has approximately

$$g_R(T) \simeq \sum_{X_b} g_{X_b} + \frac{7}{8} \sum_{X_f} g_{X_f}. \tag{13.7}$$

Therefore, the *number of ultra-relativistic degrees of freedom* is a growing function of temperature, since, depending on the value of T, different particles are in ultra-relativistic thermal equilibrium and, the higher is the temperature, the higher is the number of particle species that can be in ultra-relativistic thermal equilibrium.

For example, consider temperatures $T \gg m_e c^2/2 \simeq 0.25 \, \text{MeV}$, where m_e is the electron mass. At these temperatures, photons are so energetic that processes[4]

$$\gamma + \gamma \leftrightarrow e^+ + e^- \tag{13.8}$$

balance with each other and electrons and positrons can be considered in ultra-relativistic thermal equilibrium. The value of the number of ultra-relativistic degrees of freedom for the gas of photons, electrons and positrons is then given by

$$g_R^{\gamma + e^+ + e^-} (T \gg m_e c^2/2) = g_\gamma + \frac{7}{8} (g_{e^-} + g_{e^+}) = \frac{11}{2}, \tag{13.9}$$

where we used $g_\gamma = g_{e^-} = g_{e^+} = 2$. Notice that the electric charge neutrality imposes now, more generally, the condition on the number densities

$$n_{e^-} - n_{e^+} = n_p. \tag{13.10}$$

When $T \ll m_e c^2/2$, the positron number density becomes negligible and one recovers the condition $n_{e^-} = n_p$.

---

[3]Photons and neutrinos do not obey this rule. Photons are bosons with spin $S = 1$ but $g_\gamma = 2$ instead of 3 and neutrinos are fermions with spin $S = 1/2$ but $g_\nu = 1$ instead of 2.

[4]Photons producing pairs of particles and anti-particles are called *pair production* processes, while the inverse processes are called *annihilations*.

13.2 NEUTRINO DECOUPLING

What other particle species contribute to g_R if the thermal energy is higher than their mass rest energy? We have to use our knowledge of particle physics, currently nicely encoded in the standard model. Let us concentrate on that stage of the early universe history when $T \lesssim 10\,\mathrm{MeV}$. The masses of pions ($\pi$), muons ($\mu$) and tauons ($\tau$) are much higher than $10\,\mathrm{MeV}/c^2$ and, therefore, they do not contribute to $g_R(T)$.[5] However, one has also to consider three species of neutrinos, one for each family: (i) electron-neutrinos ν_e, (ii) muon-neutrinos ν_μ, (iii) tauon-neutrinos ν_τ. From tritium β-decay experiments [1], combined with neutrino mixing experiments [2], we know that the three neutrino masses satisfy the upper bound (at 95% C.L.)[6]

$$m_{\nu_i} \lesssim 2\,\mathrm{eV}/c^2 \,. \tag{13.11}$$

This implies that neutrinos could potentially contribute to $g_R(T)$. Why *potentially*? Because one has to verify whether neutrinos are indeed in thermal equilibrium [3]. This can be done proceeding analogously to what we did when we calculated the condition for the thermalisation of photons mediated by Thomson scattering on electrons. In the case of neutrinos, thermal equilibrium holds if

$$\Gamma_\nu \gtrsim H \,, \tag{13.12}$$

where Γ_ν is the total interaction rate of neutrinos. Neutrinos can interact with photons and electrons through various processes, such as neutrino pair production and annihilations,

$$e^+ + e^- \leftrightarrow \nu_e + \bar{\nu}_e \,, \tag{13.13}$$

that are able to change the number of neutrinos.[7] Neutrinos interact only through *weak interactions* and, therefore, their interaction rate is much lower than the electron-photon interaction rate governed by electromagnetic interactions. It is found that

$$\Gamma_\nu \simeq \frac{2}{\hbar} \frac{G_F^2\,T^5}{(\hbar c)^6} \,, \tag{13.14}$$

---

[5]The up and down quarks and anti-quarks, that would be lighter than $10\,\mathrm{MeV}$, at these temperatures are not free but bound in pions. Quarks become free only at temperatures above the temperature of the quark-hadron phase transition $T_{\mathrm{qh}} \simeq 150\,\mathrm{MeV}$.

[6]We have seen that from CMB anisotropies combined with external information a stringent one derives a stringent upper bound on the sum of neutrino masses $\sum_i m_{\nu_i} \lesssim 0.194\,\mathrm{eV}$ (at 95% C.L.). When this is combined with the information from neutrino mixing experiments, it translates into an upper bound $m_{\nu_i} \lesssim 0.065\,\mathrm{eV}$ on each neutrino mass that is of course much more stringent than (13.11). However, the upper bound from Tritium β-decay experiments is model independent and in any case it is sufficient for our considerations.

[7]These are *inelastic reactions*. One has also to consider *elastic* reactions, such as $e^- + \nu_e \leftrightarrow e^- + \nu_e$, where the same particles are in the initial and in final state. Even though they cannot change the number of neutrinos, they are able to distribute energy and momentum contributing to the establishment of *kinetic* equilibrium. Inelastic processes contribute not only to establishing kinetic equilibrium, but also *chemical* equilibrium for the number of neutrinos. Thermal equilibrium is realised when both kinetic and chemical equilibrium are achieved.

where $G_F/(\hbar c)^3 \simeq 10^{-5}\,\mathrm{GeV}^{-2}$ is the Fermi constant. From this expression and from Eq. (13.3), one finds

$$\Gamma_\nu \gtrsim H \quad \Leftrightarrow \quad T \gtrsim T^\nu_{\mathrm{dec}} \simeq 1\,\mathrm{MeV}\,. \tag{13.15}$$

This temperature is the analogue of the matter-radiation decoupling temperature. Therefore, analogously to photons, below this temperature neutrinos can freely propagate. If the initial temperature of the radiation-dominated regime is higher than T^ν_{dec}, neutrinos could be produced and reach thermal equilibrium in the early universe. In this case there should exist today a *relic neutrino background*, in the same way as a CMB exists. However, relic neutrinos interact so weakly that nobody has yet found a way to detect them.

If, one day, the thermal relic neutrino background will be discovered, it will provide direct proof of the occurrence of a hot stage in the early universe at temperatures much above $\sim 1\,\mathrm{MeV}$. However, even in the absence of direct evidence, we have two strong reasons to believe that such high temperatures were indeed reached in the early universe. The first one is that the acoustic peaks in the CMB temperature anisotropies power spectrum are also sensitive to the number of ultra-relativistic degrees of freedom during recombination [4]. From their fit, one finds (see again Table 12.2) that the so-called *effective number of neutrino species*[8] contributing to g_R during recombination is given by [5]

$$N^{\mathrm{dec}}_\nu = 3.13^{+0.64}_{-0.63}\,. \tag{13.16}$$

This result excludes a vanishing value with great statistical significance and, at the same time, it supports the standard model expectation $N^{\mathrm{dec}}_\nu = 3$. A second strong reason to believe that neutrinos were indeed produced in thermal equilibrium in the early universe, is provided by *Big Bang nucleosynthesis*, that we are going to discuss in the next chapter.

In light of these two arguments, we can then calculate the number of ultra-relativistic degrees of freedom in the standard model at temperatures $T \gtrsim 1\,\mathrm{MeV}$, finding[9]

$$g_R(T \gtrsim 1\,\mathrm{MeV}) = g^{\gamma + e^\pm + 3\nu}_R = g_\gamma + \frac{7}{8}\left[g_{e^-} + g_{e^+} + 3 \times (g_\nu + g_{\bar\nu})\right] = \frac{43}{4} = 10.75\,, \tag{13.17}$$

where we used $g_\nu = g_{\bar\nu} = 1$.[10]

---

[8]We will carefully define it in a moment.

[9]This expression is valid up to temperatures $T \simeq m_\mu \simeq (m_\mu/2)\,\mathrm{MeV}/c^2 \simeq 50\,\mathrm{MeV}$, where m_μ is the muon mass.

[10]Neutrinos, are fermions with spin $S = 1/2$ and so, as for electrons and positrons, one would expect $g_\nu = g_{\bar\nu} = 2$. However, they are somehow special and in the standard model they exist only as quantum states with *left-handed* helicity and there are no neutrino quantum states with *right-handed* helicity. Vice versa anti-neutrino quantum states exist only with right-handed helicity and left-handed anti-neutrino quantum states do not exist. Helicity is the projection of spin along the direction of motion.

Once T drops below $1\,\mathrm{MeV}$, the value of $g_R(T)$ remains approximately constant down to $T \simeq m_e c^2 \simeq 0.5\,\mathrm{MeV}$. At this time electrons and positrons start to exit the ultra-relativistic regime since annihilation and pair production processes do not balance any more. Pair production processes start to become more and more rare compared to annihilations and the number densities of electrons and positrons get Boltzmann suppressed, $n_{e^{\pm}} \propto \exp[-m_e c^2/T]$. At the time when $T \simeq m_e c^2/3 \simeq 0.15\,\mathrm{MeV}$, positrons have almost completely disappeared, while the initial small excess of non-relativistic electrons survives thanks to electric charge neutrality and eventually the limit $n_{e^-} = n_p$ is recovered (see Exercise 13.4).

What about neutrinos? Even though they are decoupled, they are still ultra-relativistic and continue to contribute to g_R. Naively, one could then think that $g_R(T \ll m_e c^2) \simeq g_R^{\gamma+3\nu's} = 2 + 42/8 = 29/4 = 7.25$. However, this result neglects a subtle effect that has to be taken into account. When electrons and positrons annihilate, they annihilate only into photons, since processes able to transfer energy (and entropy!) to neutrinos, like those in Eq. (13.13), are *switched off*. Therefore, electron-positron annihilations transfer energy from electrons and positrons only to photons but not to neutrinos. Because of this effect, it can be shown that the neutrino-to-photon temperature ratio, for $T \ll m_e c^2/2 \simeq 0.25\,\mathrm{MeV}$, is given by $T_{\nu,0}/T_0 \simeq (4/11)^{1/3} \simeq 0.7138$.[11] Correspondingly, the number of ultra-relativistic degrees of freedom has to be correctly calculated using

$$g_R(T \ll 0.25\,\mathrm{MeV}) \simeq g_R^{\mathrm{dec}} = 2 + 3 \times \frac{7}{4}\left(\frac{T_{\nu,0}}{T_0}\right)^4 \simeq 3.36\,, \tag{13.18}$$

where $g_R^{\mathrm{dec}} \equiv g_R(T_{\mathrm{dec}})$. This number would remain constant until present (though we do not live in a radiation-dominated regime anymore!) if neutrinos were massless as in the standard model.

With this result, we can finally recalculate the value of $\Omega_{R,0}$ including the contribution from neutrinos in addition to the contribution from relic photons calculated in Chapter 10 (see Eq. (10.38)), finding

$$\varepsilon_{R,0} = \varepsilon_{\gamma,0}^{CMB} + \varepsilon_{\nu,0} = g_{R,0}\frac{\pi^2}{30}\frac{T_0^4}{(\hbar c)^3} \Rightarrow \Omega_{R,0} = \Omega_{\gamma,0}^{CMB} + \Omega_{\nu,0} \simeq 0.90 \times 10^{-4}\,. \tag{13.19}$$

In this way we obtain a more precise determination of all relevant quantities at the matter-radiation equality time:

$$a_{\mathrm{eq}} \simeq 2.9 \times 10^{-4}\,, \ z_{\mathrm{eq}} \simeq 3400\,, \ T_{\mathrm{eq}} \simeq 0.82\,\mathrm{eV}\,, \ t_{\mathrm{eq}} \simeq 50,000\,\mathrm{yr}\,. \tag{13.20}$$

The account of neutrinos clearly delays the matter-radiation equality time.

The Eq. (13.18) gives the value of g_R^{dec} in the case of the standard model, containing three neutrino species. One can extend the calculation of g_R^{dec} in models beyond the standard model. In this case, one typically has to consider an additional contribution to g_R^{dec} from a possible extra-amount of radiation during recombination,

---

[11]This result can be shown using entropy conservation: see Exercise 13.2.

often referred to as *dark radiation*. This is usually parameterised in terms of N_ν^{dec}, the number of effective neutrino species at decoupling[12]:

$$g_R^{\text{dec}} = 2 + N_\nu^{\text{dec}} \times \frac{7}{4} \left(\frac{T_{\nu,0}}{T_0} \right)^4 . \tag{13.21}$$

As we have seen, the acoustic peaks are sensitive to dark radiation and the *Planck* satellite has determined N_ν^{dec} quite precisely (see Eq. (13.16)). This result places quite a stringent constraint on models of new physics. In particular, it constrains models containing additional light sterile neutrino states.[13] More generally, this is a beautiful example illustrating how the physics of the early universe can be used as a laboratory of particle physics.

13.3 CALCULATION OF THE NUMBER OF ULTRA-RELATIVISTIC DEGREES OF FREEDOM IN THE STANDARD MODEL

There are no compelling experimental evidences that the initial temperature of the radiation-dominated regime, T_{in}, is higher than $\sim 10\,\text{MeV}$. However, there are strong theoretical reasons to believe that T_{in} was in fact even higher than $m_t c^2 / 2 \sim 100\,\text{GeV}$, where m_t is the mass of the top quark, the heaviest among the standard model elementary particles. For this reason, it is useful to calculate the value of $g_R(T)$ up to values of the temperature as high as $\sim 100\,\text{GeV}$, corresponding to a thermalisation of all standard model degrees of freedom.

A full calculation is shown in Table 13.1. One can see that the maximum value of $g_R(T)$ in the standard model is given by $g_R(T \gtrsim 100\,\text{GeV}) = g_{SM} = 106.75$. Let us shortly describe the jumps in the value of $g_R(T)$ at temperatures above $\sim 10\,\text{MeV}$ due to the thermalisation of particle species heavier than electrons and positrons. The first jump occurs when muons are produced at $T \simeq m_\mu c^2 / 2 \simeq 50\,\text{MeV}$. At this temperature one has

$$g_R(T \simeq 50\,\text{MeV}) \simeq g_R^{\gamma + e^\pm + 3\nu} + g_R^{\mu^\pm} = \frac{57}{4} = 14.25 . \tag{13.22}$$

This value remains approximately constant up to temperatures $T \simeq m_\pi c^2 / 2 \simeq 75\,\text{MeV}$, when pions $(\pi^\pm, \pi^0)$ thermalise.[14] Above this temperature, $g_R(T)$ has to

---

[12]There is an even more subtle effect yielding another twist in the story: even in the standard model one has $N_\nu^{\text{dec}} \neq 3$! This happens because it is more correct to say that electrons and positrons annihilate *almost* only into photons. In fact there is a very small fraction of *non-thermal neutrinos* produced by electron-positron annihilations after neutrino decoupling. In this way the number of effective neutrino species at decoupling, as defined in Eq. (13.21), even in the standard model, is not exactly 3 but more precisely $(N_\nu^{\text{dec}})_{SM} = 3.046$ [6]. This should clearly illustrate why it is more correct to talk of *effective* number of neutrinos. This small effect slightly increases the radiation contribution further delaying the matter-radiation equality time. With this correction, one has: $g_R^{\text{dec}} \simeq 3.385$, $\Omega_{R,0} \simeq 0.915 \times 10^{-4}$ and $t_{\text{eq}} \simeq 52,000\,\text{yr}$ (the last two are also the figures reported in the table of constants at the end of the book).

[13]For example, it rules out the case of one fully thermalised light sterile neutrino species predicted by solutions to the anomalous neutrino oscillation experimental results found in the LSND experiment [7, 8].

[14]Pions are particles with spin 0 made of 1 up or down quark and 1 up or down anti-quark.

TABLE 13.1 Dependence of g_R on temperature in the standard model.

T	g_R	Particle content
$m_e c^2/2 \simeq 0.25\,\mathrm{MeV} \gg T \geq T_0$	3.36	$\gamma + 3$ massless $\nu's$
$m_\mu c^2/2 \simeq 50\,\mathrm{MeV} \gg T \gg m_e c^2/2$	$43/4 =10.75$	$\ldots + e^\pm$
$m_\pi c^2/2 \simeq 75\,\mathrm{MeV} \gg T \gg m_\mu c^2/2$	$57/4 =14.25$	$\ldots + \mu^\pm$
$T_{\mathrm{qh}} \simeq 150\,\mathrm{MeV} \gg T \gg m_\pi c^2/2$	$69/4 =17.25$	$\ldots + \pi^0, \pi^\pm$
$m_\tau c^2/2 \gtrsim m_c c^2/2 \simeq 0.65\,\mathrm{GeV} \gg T \gtrsim T_{\mathrm{qh}}$	61.75	$\ldots + u,d,s$ quarks $+$ 8 gluons
$m_b c^2/2 \simeq 2\,\mathrm{GeV} \gg T \gg m_\tau c^2/2$	75.75	$\ldots + \tau^\pm + c$ quark
$m_{W,Z,H^0} c^2/2 \simeq 40\,\mathrm{GeV} \gg T \gg m_b c^2/2$	86.25	$\ldots + b$ quark
$m_t c^2/2 \simeq 90\,\mathrm{GeV} \gg T \gg m_{W,Z,H^0} c^2/2$	96.25	$\ldots + W^\pm, Z^0, H^0$ bosons
$T \gg m_t c^2/2$	106.75	$\ldots +$ top quark

include the contribution from pions, $g_R^{\pi^\pm+\pi^0} = 3$, so that $g_R(T \simeq 75\,\mathrm{MeV}) = 69/4$. This value is valid up to $T_{\mathrm{qh}} \simeq 150\,\mathrm{MeV}$, the temperature of the so-called *quark-hadron phase transition*. Above this temperature, number densities become so high that, because of a genuine effect of *quantum chromodynamics* called *asymptotic freedom*, quarks in pions become unbound and the degrees of freedom of up, down and strange quarks (and anti-quarks) plus those of the 8 gluons give a contribution $g_R^{u,d,s+8\,\mathrm{gluons}} = 47.5$ to $g_R(T)$. In this way, one obtains

$$g_R(T \simeq 150\,\mathrm{MeV}) = g_R^{\gamma+e^\pm+3\nu} + g_R^{\mu^\pm} + g_R^{u,d,s+8\,\mathrm{gluons}} = 61.75 . \qquad (13.23)$$

We refer to Table 13.1 for values of g_R at higher temperatures, up to a maximum value $g_{SM} = 106.75$, saturated for $T \gtrsim 100\,\mathrm{GeV}$.

As we already noticed discussing the value of g_R during recombination, in extensions of the standard model one expects modifications to $g_R(T)$. These are typically coming from the existence of new degrees of freedom, and can have an impact on certain cosmological observables. We discussed already how the acoustic peaks in CMB temperature anisotropies are sensitive to a modification of g_R during recombination. In the next chapter we will discuss another important (and historical) set of cosmological observables sensitive to the value of g_R at earlier times: the primordial nuclear abundances.

EXERCISES

Exercise 13.1 *Assuming thermal equilibrium, show that the entropy of particles contained in a portion of comoving volume is conserved.*

Exercise 13.2 *Using entropy conservation, show that the temperature of the neutrino background at present is given by $T_\nu = T_0 \left(4/11\right)^{1/3}$.*

Exercise 13.3 *Calculate how the value of g_R change at temperatures $m_\mu\, c^2/2 \simeq 50\,\text{MeV} \gg T \gg m_e\, c^2/2 \simeq 0.25\,\text{MeV}$ assuming that right-handed neutrinos and left-handed anti-neutrinos exist and are in ultra-relativistic thermal equilibrium.*

Exercise 13.4 *Derive an exact expression, in integral form, for the number of ultra-relativistic degrees of freedom of electrons and positrons as a function of temperature, describing the transition from ultra-relativistic to non-relativistic equilibrium in a continuous way while temperature decreases. Derive also an expression, always in integral form, for the number densities of electrons and positrons as a function of temperature. Finally, using the results of Exercise 13.2, calculate also the evolution of T_ν as a function of temperature. Plot the obtained expressions.*

Exercise 13.5 *Calculate the matter-radiation equality time including the contribution from neutrinos and confirming the result $t_{\text{eq}}^{RM} \simeq 50,000\,\text{yr}$ (see Eq. (13.20)).*

BIBLIOGRAPHY

[1] C. Kraus, B. Bornschein, L. Bornschein, J. Bonn, B. Flatt, A. Kovalik, B. Ostrick and E. W. Otten et al., *Final results from phase II of the Mainz neutrino mass search in tritium beta decay*, Eur. Phys. J. C **40** (2005) 447; V. N. Aseev et al. [Troitsk Collaboration], *An upper limit on electron antineutrino mass from Troitsk experiment*, Phys. Rev. D **84** (2011) 112003. E. W. Otten and C. Weinheimer, *Neutrino mass limit from tritium beta decay*, Rept. Prog. Phys. **71** (2008) 086201; K. A. Olive et al. [Particle Data Group Collaboration], *Review of particle physics. Section 14: Neutrino mass, mixing and oscillations*, Chin. Phys. C **38** (2014) 090001.

[2] C. Patrignani et al. [Particle Data Group], *Review of particle physics*, Chin. Phys. C **40** (2016) no.10, 100001.

[3] For specialistic essays on neutrinos in cosmology see: A. D. Dolgov, *Neutrinos in cosmology*, Phys. Rept. **370** (2002) 333 [hep-ph/0202122]; J. Lesgourgues, G. Mangano, G. Miele, S. Pastor, *Neutrino cosmology*, Cambridge, Cambridge University Press, 2013.

[4] S. Hannestad and G. Raffelt, *Imprint of sterile neutrinos in the cosmic microwave background radiation*, Phys. Rev. D **59** (1999) 043001 [astro-ph/9805223].

[5] P. A. R. Ade et al. [Planck Collaboration], *Planck 2015 results. XIII. Cosmological parameters*, Astron. Astrophys. **594** (2016) A13 [arXiv:1502.01589].

[6] R. E. Lopez, S. Dodelson, A. Heckler and M. S. Turner, *Precision detection of the cosmic neutrino background*, Phys. Rev. Lett. **82** (1999) 3952 [astro-ph/9803095]; G. Mangano, G. Miele, S. Pastor, T. Pinto, O. Pisanti and P. D. Serpico, *Relic neutrino decoupling including flavor oscillations*, Nucl. Phys. B **729** (2005) 221 [hep-ph/0506164].

[7] A. Aguilar-Arevalo et al. [LSND Collaboration], *Evidence for neutrino oscillations from the observation of anti-neutrino(electron) appearance in a anti-neutrino(muon) beam*, Phys. Rev. D **64** (2001) 112007 [hep-ex/0104049].

[8] S. M. Bilenky, C. Giunti, W. Grimus and T. Schwetz, *Four neutrino mixing and Big Bang nucleosynthesis*, Astropart. Phys. **11** (1999) 413 [hep-ph/9804421]; P. Di Bari, *Update on neutrino mixing in the early universe*, Phys. Rev. D **65** (2002) 043509 [hep-ph/0108182].

CHAPTER 14

Big Bang nucleosynthesis

A s discussed in Chapter 2, the primordial nuclear composition of the universe is one of the cosmological pillars that a successful cosmological model needs to explain. In this chapter we show how this can be well understood within the Hot Big Bang scenario thanks to Big Bang nucleosynthesis (BBN). This model implies that the initial temperature of the early universe T_{in} had to be sufficiently high so that all nucleons were unbound and weak reactions were inter-converting neutrons into protons in equilibrium, implying $T_{\text{in}} \gg 1\,\text{MeV}$.

14.1 PRIMORDIAL NUCLEAR ABUNDANCES

Nuclear abundances are certainly synthesised in stars. In particular heavy nuclei (metals)[1] can be well explained within a conventional stellar nucleosynthesis. However, light elements, such as deuterium and ^{4}He, are much more abundant than expected from stellar nucleosynthesis.

More precisely, the astronomical observations indicate that the universe is basically composed of about 75% hydrogen-1 (protium) and 25% helium-4, plus very small traces of other light elements like hydrogen-2 (deuterium), helium-3, lithium-6 and lithium-7. Only a small fraction of these abundances can be explained by stellar nucleosynthesis.

Let us define the *nuclear abundance* X_i for the element $i \equiv (A, Z)$ with a number Z of protons (the charge number) and a total number A of nucleons (the mass number), as the fraction of nucleons contained in the nuclei of the element, explicitly

$$X_i \equiv \frac{A\,n_i}{n_N}, \qquad (14.1)$$

where n_i is the number density of the element i and n_N is the total number density of nucleons, thus $\sum_i X_i = 1$. The ^{4}He abundance[2] is traditionally indicated with Y_{p}. Astronomical observations find that the universe is in first approximation just

---

[1]In astrophysics, metals are all elements heavier than helium.
[2]The nuclei of ^{4}He are α particles composed of 2 protons and 2 neutrons so that $Z = 2$ and $A = 4$.

made of hydrogen-1, with a primordial abundance $X_p \simeq 0.75$, and ^{4}He, with a primordial abundance, $Y_p \simeq 0.25$. All other nuclear abundances are much smaller than unity.

The theory of Big Bang nucleosynthesis, based on the Hot Big Bang model, provides an elegant explanation for this observed primordial nuclear composition of the universe [1, 2, 3].

14.2 NEUTRON-TO-PROTON RATIO: THERMAL EQUILIBRIUM STAGE

At temperatures $T \gtrsim 10\,\mathrm{MeV}$ the mean energies of the elementary particles present in the primordial plasma, $\gamma + 3\nu$'s $+e^\pm$, are much higher than the nuclear binding energies per nucleon and therefore, if thermal equilibrium holds, almost all nucleons (neutrons and protons) are unbound.

Neutrinos play a very important role since they are exchanged in weak interaction particle processes able to inter-convert neutrons into protons and vice versa. These are given by

$$n \;\leftrightarrow\; p + e^- + \bar{\nu}_e \,, \tag{14.2}$$
$$n + e^+ \;\leftrightarrow\; p + \bar{\nu}_e \,, \tag{14.3}$$
$$n + \nu_e \;\leftrightarrow\; p + e^- \,. \tag{14.4}$$

The first of these processes, proceeding from left to right, is *neutron decay* and, as we will see, it plays a special role. At $T \gtrsim 10\,\mathrm{MeV}$ these processes are fast enough to enforce thermal equilibrium[3] and the number density of nucleons is then described by the Maxwell–Boltzmann distribution

$$n_N^{\mathrm{eq}} \propto m_N^{3/2}\, e^{-\frac{m_N c^2}{T}} \quad (N = p, n)\,. \tag{14.5}$$

The *neutron-to-proton ratio* $n/p \equiv n_n/n_p$ is then well approximated by its thermal equilibrium limit[4]

$$\frac{n}{p} \simeq \left(\frac{n}{p}\right)_{\mathrm{eq}} \simeq e^{-\frac{Q_n}{T}} \,, \tag{14.8}$$

---

[3]Therefore, we can define as BBN onset temperature $T_{BBN}^{\mathrm{onset}} \sim 10\,\mathrm{MeV}$.

[4]More precisely one has, in natural units,

$$n_N^{\mathrm{eq}} = g_N \left(\frac{m_N T}{2\pi}\right)^{\frac{3}{2}} e^{\frac{\mu_N - m_N}{T}} \quad (N = p, n)\,, \tag{14.6}$$

where $g_n = g_p = 2$ is the nucleon spin degeneracy and μ_n and μ_p are the chemical potentials of neutrons and protons. Therefore, the thermal equilibrium neutron-to-proton ratio is given by

$$\left(\frac{n}{p}\right)_{\mathrm{eq}} = \left(\frac{m_n}{m_p}\right)^{\frac{3}{2}} e^{\frac{\mu_n - \mu_p}{T}} e^{-\frac{Q_n}{T}} \,. \tag{14.7}$$

Because of the thermal equilibrium in the reactions (14.3) and (14.4), one has the relation $\mu_n - \mu_p = \mu_{e^-} - \mu_{\nu_e}$. From different arguments we know that $(\mu_{e^-} - \mu_{\nu_e})/T \ll 1$. Moreover one can also employ the approximation $(m_n/m_p)^{3/2} \simeq 1$. These two approximations justify the second approximation step in Eq. (14.8).

where $Q_n \equiv (m_n - m_p)c^2 \simeq 1.29\,\text{MeV}$ is the difference between the neutron and the proton rest energies. For $T \gg Q_n$ one has $n/p \simeq 1$, but when T drops below Q_n, the Boltzmann factor suppresses the neutron abundance and, if thermal equilibrium continued to hold, there would be eventually only protons left for $T \ll Q_n$ and no neutrons. We would therefore live in a universe with only hydrogen-1 atoms. However, the observations show that there is also a sizeable helium-4 primordial abundance. The validity of the thermal equilibrium approximation has therefore to break down at some stage during the expansion.

14.3 NEUTRON-TO-PROTON RATIO: FREEZE-OUT

The neutron-to-proton ratio is well described by the thermal equilibrium limit Eq. (14.8) only when the condition

$$\Gamma_{n\leftrightarrow p}(T) \simeq \frac{2}{\hbar} \frac{G_F^2}{(\hbar c)^6} T^5 \xi(T) \gtrsim H(T) \tag{14.9}$$

is verified, where $\Gamma_{n\leftrightarrow p}$ is the rate of the $2 \leftrightarrow 2$ scattering reactions interconverting neutrons and protons (see Eqs. (14.3) and (14.4)) and $\xi(T)$ is a function of temperature that in the limit $T \gg Q_n$ tends to unity while it is greater for $T \lesssim Q_n$ (see Exercise 14.1). Notice that this rate is (not by chance) approximately similar to the total rate of neutrino interactions, since they are indeed both determined by weak interactions. Notice also that this rate does not include neutron decay since this is a special process that has to be discussed separately.

Using Eq. (13.3) for H, the condition (14.9) translates into a condition on the temperature given by

$$T \gtrsim T_f \simeq \left[\frac{1}{2\,\xi(T_f)} \sqrt{\frac{8\pi^3\,g_R(T)}{90}} \frac{(\hbar c)^6/G_F^2}{M_P c^2} \right]^{\frac{1}{3}} \simeq 0.8\,\text{MeV}, \tag{14.10}$$

where T_f is the *freeze-out* temperature for the neutron-to-proton ratio and we used $g_R = 10.75$ and the result $\xi^{-1}(T_f) \simeq 3$ (see Exercise (14.1)). Using this value of the freeze-out temperature, one can then calculate the neutron-to-proton ratio at freeze-out,

$$\left(\frac{n}{p}\right)_f = e^{-\frac{Q_n}{T_f}} \simeq 0.20. \tag{14.11}$$

Moreover from the time-temperature relation Eq. (13.5) one finds for age of the universe at freeze-out

$$t_f \simeq 1.0\,\text{s}. \tag{14.12}$$

However, there is a process that does not freeze out and and that we still have to take into account: *neutron decay.*[5]

---

[5]On the contrary, one can say that it rather *freezes in*, since the ratio of the neutron decay rate to the expansion rate, $\Gamma_n/H \sim t/\tau_n$ increases with time and at some point the age of the universe

14.4 NUCLEOSYNTHESIS

If nothing intervenes, when the age of the universe t becomes much higher than the neutron life-time τ_n, all neutrons would decay into protons and the large observed helium-4 abundance could not be explained. However, before neutrons have fully decayed, the formation of nuclei from fusion processes, nucleosynthesis, occurs.

The discussion proceeds along similar lines for the case of recombination. The difference is that in the case of nucleosynthesis there is a competition among different nuclear abundances. The lightest stable nucleus is deuterium, with a binding nuclear energy per nucleon given by $(B/A)_D \simeq 1.11\,\mathrm{MeV}$. However, as in the case of recombination, the formation is postponed because of the large number of photons per nucleon able to photo-disintegrate nuclei, in particular deuterium through the reaction

$$\gamma + \mathrm{D} \to n + p\,. \tag{14.13}$$

If one assumes thermal equilibrium as a good description for the synthesis of nuclear abundances, one obtains, as in the case of recombination, a Saha equation for each nuclear element i,

$$X_i^{\mathrm{eq}} = g_A\, C_A\, A^{\frac{5}{2}} \left(\frac{T}{m_N}\right)^{\frac{3(A-1)}{2}} \eta_{B,0}^{A-1}\, X_p^Z\, X_n^{A-Z}\, e^{\frac{B_i}{T}}\,, \tag{14.14}$$

where g_A is the nuclear spin degeneracy, $m_N \simeq m_p \simeq m_n$ is the nucleon mass, $B_i = [Z\, m_p + (A-Z)\, m_n - m_A]\, c^2$ is the nuclear binding energy and we defined

$$C_A = \zeta(3)^{A-1}\, \pi^{\frac{1}{2}(1-A)}\, 2^{\frac{3A-5}{2}}\,. \tag{14.15}$$

Analogously to the definition of recombination temperature, we can now define a *nucleosynthesis temperature* T_{nuc}^i for the element i such that

$$\frac{X_i(T_{\mathrm{nuc}}^i)}{X_p^Z\, X_n^{A-Z}} \equiv 1\,. \tag{14.16}$$

From Eq. (14.14) it is also possible to derive an approximate expression for T_{nuc}^i valid for thermal equilibrium,

$$(T_{\mathrm{nuc}}^i)_{\mathrm{eq}} \simeq \frac{B_A/(A-1)}{39 - (A-1)\ln(\eta_{B,0}/10^{-10}) - \ln(C_A)/(A-1)}\,. \tag{14.17}$$

becomes larger than the neutron life-time given by the inverse of the decay rate at zero temperature (the neutron decay width). Therefore, the freeze-out condition applies only to scatterings, typically two particle scatterings, even though sometimes scatterings with more than two particles in the initial state can be important and have to be taken into account. In the case of particle decays in general one has also to consider the effect of time dilation when the thermal energy is much higher than the rest energy of the decaying particle, since in this case these will move (on average) ultra-relativistically with respect to the comoving system. This means that if τ_X is the life time at rest of a particle X, the decay rate $\Gamma_X = (\langle\gamma\rangle\,\tau_X)^{-1}$, where $\langle\gamma\rangle$ is the thermally averaged Lorentz factor. However, in the case of neutrons, with a mass $m_n \sim 1\,\mathrm{GeV}/c^2$, their life time at rest is so long, $\tau_n \simeq 880\,\mathrm{s}$ [4], that in any case they start to decay only when $T \ll m_n\, c^2$ and time dilation can be neglected, i.e., $\langle\gamma\rangle \simeq 1$.

However, it is easy to understand that these results, assuming thermal equilibrium, do not provide the correct picture but only some partial information that has to be complemented. First of all notice that in the expression (14.14) the nuclear binding energy is roughly linear in A. Given a neutron abundance X_n and a proton abundance X_p, it clearly predicts that, for low enough temperatures, heavy elements should be by far more abundant than light ones (excluding hydrogen-1). This is clearly in disagreement with the measured primordial nuclear abundances showing that the ^{4}He abundance dominates by many orders of magnitude over the abundances of heavier elements.

As we have already seen in the case of the calculation of the neutron-to-proton ratio, the problem with thermal equilibrium is that it assumes that there is always some sufficiently fast process able to bring the system into equilibrium. It neglects whether the processes, necessary for the establishment of thermal equilibrium, are effectively in operation or not. In the case of nucleosynthesis the failure of such an assumption is even more blatant. Thermal equilibrium predicts ^{4}He to be the first element to form since, using Eq. (14.17), it would have the highest nucleosynthesis temperature $T_{\mathrm{nuc}}^{^4\mathrm{He}} \simeq 0.28\,\mathrm{MeV}$.[6] The problem is that the synthesis of ^{4}He proceeds through the reactions

$$D + n \quad \to \quad T + \gamma\,, \tag{14.18}$$
$$T + D \quad \to \quad n + {}^4\,He\,, \tag{14.19}$$

where T indicates tritium with $A = 3$ and $Z = 1$ (i.e., 1 proton and 2 neutrons). This means that the ^{4}He-synthesis requires the presence of deuterium and, therefore, it cannot start prior to deuterium synthesis. This is also true for all the other nuclear abundances and for this reason deuterium, contrarily to what is expected from thermal equilibrium, is the first nuclear element to be synthesised at a temperature $T_{\mathrm{nuc}} \simeq 0.065\,\mathrm{MeV}$, again calculated using Eq. (14.17). At least in this case thermal equilibrium does give the correct answer. At this time, deuterium-synthesis can effectively proceed through the fusion reaction

$$p + n \to D + \gamma\,. \tag{14.20}$$

All neutrons get very quickly trapped into deuterium nuclei and neutron decays get inhibited.

Using the time-temperature relation Eq. (13.5) with $g_R = 3.36$ (cf. Eq. (13.18)), one finds that the age of the universe at deuterium-synthesis is approximately

$$t_{\mathrm{nuc}} \simeq 310\,s\,. \tag{14.21}$$

The decrease of the ratio (n_n/n_p) from the freeze-out to the nucleosynthesis is simply described by the decaying law

$$\frac{(n/p)_{\mathrm{nuc}}}{(n/p)_{\mathrm{f}}} = e^{-\frac{t_{\mathrm{nuc}}}{\tau_n}} \simeq e^{-\frac{310}{880}} \simeq 0.7\,, \tag{14.22}$$

---

[6]We adopt here, like also below for the case of deuterium, the value from CMB for $\eta_{B,0}$ (see Eq. (12.40)).

so that

$$(n/p)_{\mathrm{nuc}} \simeq 0.15 \,. \tag{14.23}$$

Once that deuterium-synthesis§ starts, the ^{4}He-synthesis and, to a much lower extent of the heavier nuclei, proceeds very quickly since it is mediated by strong interactions. For this reason t_{nuc} can be considered the time of nucleosynthesis in general, not just for deuterium. Therefore, the formation of the ^{4}He abundance, that according to thermal equilibrium should occur already when $T_{\mathrm{nuc}}^{^4\mathrm{He}} \simeq 0.28\,\mathrm{MeV}$, is delayed by the fact that it cannot proceed until deuterium forms.[7]

Because of the much higher binding energy per nucleon of ^{4}He compared to the other nuclei of comparable mass number, this is by far the most favoured nuclear element and one can say that with very good approximation all neutrons eventually end up forming ^{4}He nuclei. The formation of very heavy elements, even more stable than ^{4}He, does not occur, as thermal equilibrium would predict, because the nuclear rates for the formation of heavier elements are much smaller than the expansion rate and the nuclear abundances freeze out, similarly to what we have seen for the neutron-to-proton ratio.[8] Therefore, the ^{4}He abundance is easily related to $(n/p)_{\mathrm{nuc}}$ by

$$Y_{\mathrm{p}} = 2\,\frac{(n/p)_{\mathrm{nuc}}}{1 + (n/p)_{\mathrm{nuc}}} \simeq 0.2456 \,. \tag{14.24}$$

This approximate theoretical prediction matches quite well the experimental observations. In Table 14.1 we show a more complete set of the relevant nuclear reactions for BBN involving light elements. For heavier elements the primordial production is negligible compared to one in stellar environments. In particular, from the reactions

$$\begin{aligned}
^4\mathrm{He} + \mathrm{D} &\rightarrow\ ^6\mathrm{Li} + \gamma \,, &(14.25)\\
^4\mathrm{He} + {}^3\mathrm{He} &\rightarrow\ ^7\mathrm{Li} + \gamma \,, &(14.26)\\
^4\mathrm{He} + {}^3\mathrm{He} &\rightarrow\ ^7\mathrm{Be} + \gamma \,, &(14.27)\\
&&(14.28)
\end{aligned}$$

small traces of ^{6}Li , ^{7}Li and ^{7}Be also form.[9] A precise calculation of all primordial nuclear abundances has to proceed numerically by solving a set of kinetic equations for all relevant abundances and the inclusion of different subtle effects [7].

14.5   DEUTERIUM AS A BARYOMETER

The final values of the nuclear abundances depend on the value of $\eta_{B,0}$. The thermal equilibrium prediction Eq. (14.14) shows explicitly such a dependence but, as we commented, it gives incorrect results. The correct predictions have to be calculated

---

[7]This is often referred to as the *deuterium bottleneck.*

[8]A crucial point to understand is why nuclear rates for the formation of elements heavier than ^{4}He are very suppressed. The reason is that there is no stable nucleus with $A = 5$. This is an even narrower bottleneck than deuterium.

[9]However, ^{7}Be later decays into ^{7}Li by electron capture during recombination [6].

TABLE 14.1   Relevant nuclear reactions for a correct computation of the light primordial nuclear abundances.

1)	$p + n \leftrightarrow D + \gamma$
2)	$D + n \leftrightarrow T + \gamma$
3)	$^3\text{He} + n \leftrightarrow {}^4\text{He} + \gamma$
4)	$^6\text{Li} + n \leftrightarrow {}^7\text{Li} + \gamma$
5)	$^3\text{He} + n \leftrightarrow T + p$
6)	$^7\text{Be} + n \leftrightarrow {}^7\text{Li} + p$
7)	$^7\text{Li} + n \leftrightarrow {}^3\text{He} + {}^4\text{He} + \gamma$
8)	$^7\text{Be} + n \leftrightarrow {}^4\text{He} + {}^4\text{He}$
9)	$D + p \leftrightarrow {}^3\text{He} + \gamma$
10)	$T + p \leftrightarrow {}^4\text{He} + \gamma$
11)	$^6\text{Li} + p \leftrightarrow {}^7\text{Be} + \gamma$
12)	$^7\text{Li} + p \leftrightarrow {}^4\text{He} + {}^4\text{He}^4$
13)	$D + {}^4\text{He} \leftrightarrow {}^6\text{Li} + \gamma$
14)	$T + {}^4\text{He} \leftrightarrow {}^7\text{Li} + \gamma$
15)	$^3\text{He} + {}^4\text{He} \leftrightarrow {}^7\text{Be} + \gamma$
16)	$D + D \leftrightarrow {}^3\text{He} + n$
17)	$D + D \leftrightarrow T + p$
18)	$D + T \leftrightarrow {}^4\text{He} + p$
19)	$D + {}^3\text{He} \leftrightarrow {}^4\text{He} + n$
20)	$^3\text{He} + {}^3\text{He} \leftrightarrow {}^4\text{He} + p + p$
21)	$D + {}^7\text{Li} \leftrightarrow {}^4\text{He} + {}^4\text{He} + n$
22)	$D + {}^7\text{Be} \leftrightarrow {}^4\text{He} + {}^4\text{He} + p$

Notice how, directly or indirectly, no abundance can be created if deuterium has not first been synthesised.

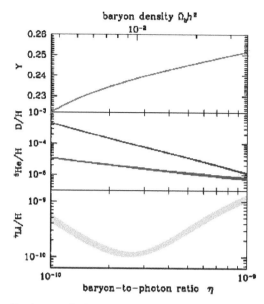

FIGURE 14.1: Predictions of the primordial nuclear abundances from Big Bang nucleosynthesis as a function of $\eta_{B,0}$ (from [5]).

solving a set of coupled kinetic equations describing the evolution of the nuclear abundances with time and taking into account the value of the rates of nuclear reactions and of course of the expansion rate. The solution is numerical and in Fig. 14.1 we show the result for the final nuclear abundances of ^{4}He, ^{3}He, D, ^{7}Li as a function of $\eta_{B,0}$ [5]. One can notice that deuterium has the strongest dependence. A quite good power-law fit to the numerical result is given by

$$(D/H)^{\mathrm{BBN}}(\eta_{B,0}) \simeq 2.6 \times 10^{-5} \left(\frac{\eta_{B,0}}{6 \times 10^{-10}}\right)^{-1.6}, \qquad (14.29)$$

having labelled, as customary, the deuterium-to-hydrogen abundance ratio with (D/H). This monotonic decrease with increasing $\eta_{B,0}$ can be qualitatively understood considering that the final abundance of deuterium is a leftover from the initial deuterium abundance synthesised at t_{nuc} and then converted into ^{4}He by the reaction (14.19). If $\eta_{B,0}$ increases, then T^D_{nuc} also increases, as one can see from Eq. (14.17), and t_{nuc} decreases. At earlier times, the rate of the reaction (14.19) is higher and, therefore, the conversion of deuterium into ^{4}He is more efficient resulting in a lower relic deuterium abundance.

This strong dependence of (D/H) on $\eta_{B,0}$ can be used for a determination of the value of $\eta_{B,0}$ alternative to that from CMB. Comparing the theoretical prediction BBN prediction Eq. (14.29) with the experimental value from astronomical observations of far quasars [8]

$$(D/H)^{\mathrm{exp}} = (2.53 \pm 0.04) \times 10^{-5}, \qquad (14.30)$$

one finds indeed for $\eta_{B,0}$ [8]

$$\eta_{B,0}^{BBN} = (6.02 \pm 0.01) \times 10^{-10}, \tag{14.31}$$

in excellent agreement with the determination from the CMB acoustic peaks Eq. (12.40). This shows how the deuterium is a fine baryometer in combination with the BBN prediction [1].

14.6 BBN BOUND ON THE NUMBER OF EFFECTIVE NEUTRINO SPECIES

The nuclear abundances depend, in addition to $\eta_{B,0}$, also on the expansion rate of the universe during nucleosynthesis determined by the number of ultra-relativistic degrees of freedom g_R. The discussion has to be done separately for ^{4}He and for deuterium, since the former is sensitive to the expansion rate at the freeze-out of (n/p), while the latter is sensitive to the expansion rate at $t_{\rm nuc}$.

In both cases it is still possible to parameterise the modifications induced by the new physics beyond the standard model in terms of the effective number of neutrino species N_ν. We can extend its definition, given at the time of decoupling in Eq. (13.21), at an arbitrary time, writing[10]

$$g_R(T) = g_R^{SM}(T) + \frac{7}{8}\left[N_\nu(T) - 3\right] \left(\frac{T_\nu(T)}{T}\right)^4. \tag{14.32}$$

A good fit of the numerical results for the ^{4}He abundance dependence on N_ν at the n/p freeze-out is provided by

$$Y_p(\eta_{B,0}, N_\nu^{\rm f}) \simeq 0.2477 + 0.0137\left(N_\nu^{\rm f} - 3\right) + 0.01 \ln\left(\frac{\eta_{B,0}}{6 \times 10^{-10}}\right), \tag{14.33}$$

where we have also indicated the slight logarithmic dependence on $\eta_{B,0}$. On the other hand the deuterium abundance dependence on N_ν at the time of nucleosynthesis is well fitted by [9].

$$(D/H)(\eta_{B,0}, N_\nu^{\rm nuc}) \simeq (D/H)^{\rm BBN}(\eta_{B,0}) \times \left[1 + 0.135\left(N_\nu^{\rm nuc} - 3\right)\right]^{0.8}, \tag{14.34}$$

where $(D/H)^{\rm BBN}(\eta_{B,0})$ is given by Eq. (14.29). Historically, the experimental measurement of Y_p provided the first hint of the existence of a third neutrino species and has placed the upper bound $N_\nu^{\rm f} \lesssim 7$ [10]. Nowadays we can use the value of $\eta_{B,0}$ from CMB, obtaining much more stringently $N_\nu^{\rm f} = 2.9 \pm 0.2$, showing no evidence of the presence of dark radiation at the n/p freeze-out.

The experimental determination of (D/H) also translates into an equally precise determination of $N_\nu^{\rm nuc} = 2.8 \pm 0.3$ and also in this case there is no evidence of the presence of dark radiation.

---

[10]The standard model case of course is recovered for $N_\nu(T) = 3$ and the value at recombination is recovered for $(T_\nu/T) = (T_\nu/T)_0 = (4/11)^{1/3}$, neglecting the small amount of neutrinos produced by $e^\pm$ annihilations as we discussed in the previous chapter.

An example of extension of the standard model able to produce a deviation of $N_\nu(T)$ from its standard model value $N_\nu = 3$, different at the three epochs probed by BBN+Y_p, BBN+(D/H) and CMB, would be given by a new particle species decaying with a life time comparable to $t_{\text{nuc}} \simeq 300\,\text{s}$.

However, as we have seen, currently all three determinations of N_ν are compatible with the standard model prediction and this strongly constrains the presence of new particle species in extensions of the standard model.[11]

EXERCISES

Exercise 14.1 *Derive an expression of the function $\xi(T)$ in the expression (14.9) for $\Gamma_{n \leftrightarrow p}$ and from this estimate the value of T_{f}.*

Exercise 14.2 *How does the predicted value of the helium-4 abundance change if the neutron life time is doubled?*

Exercise 14.3 *How would the predicted value of the Helium-4 abundance change if instead of 3 neutrino species one assumes the existence of four neutrino species in ultra-relativistic thermal equilibrium prior the onset of Big Bang nucleosynthesis (at $T \gtrsim 10\,\text{MeV}$)?*

BIBLIOGRAPHY

[1] G. Gamow, *Expanding universe and the origin of elements*, Phys. Rev. **70** (1946) 572;

[2] R. A. Alpher, H. Bethe and G. Gamow, *The origin of chemical elements*, Phys. Rev. **73** (1948) 803;

[3] R. A. Alpher and R. C. Herman, *On the relative abundance of the elements*, Phys. Rev. **74** (1948) 1737.

[4] C. Patrignani et al. [Particle Data Group], *Review of particle physics*, Chin. Phys. C **40** (2016) no. 10, 100001.

[5] R. H. Cyburt, B. D. Fields, K. A. Olive and T. H. Yeh, *Big Bang nucleosynthesis: 2015*, Rev. Mod. Phys. **88** (2016) 015004 [arXiv:1505.01076 [astro-ph.CO]].

[6] R. Khatri and R. A. Sunyaev, *Time of primordial Be-7 conversion into Li-7, energy release and doublet of narrow cosmological neutrino lines*, Astron. Lett. **37** (2011) 367 [arXiv:1009.3932 [astro-ph.CO]].

[7] R. V. Wagoner, *Big Bang nucleosynthesis revisited*, Astrophys. J. **179** (1973) 343; L. Kawano, FERMILAB-PUB-92-004-A. R. E. Lopez and M. S. Turner,

---

[11]It is comforting but at the same time demoralising from the point of view of theorists desperately seeking for signs of new physics.

Phys. Rev. D **59** (1999) 103502; [astro-ph/9807279]; F. Iocco, G. Mangano, G. Miele, O. Pisanti and P. D. Serpico, Phys. Rept. **472** (2009) 1 doi:10.1016/j.physrep.2009.02.002 [arXiv:0809.0631 [astro-ph]].

[8] R. Cooke, M. Pettini, R. A. Jorgenson, M. T. Murphy and C. C. Steidel, *Precision measures of the primordial abundance of deuterium*, Astrophys. J. **781** (2014) no. 1, 31 [arXiv:1308.3240 [astro-ph.CO]].

[9] P. Di Bari, *Update on neutrino mixing in the early universe*, Phys. Rev. D **65** (2002) 043509 [hep-ph/0108182].

[10] V. F. Shvartsman, *Density of relict particles with zero rest mass in the universe*, Pisma Zh. Eksp. Teor. Fiz. **9** (1969) 315 [JETP Lett. **9** (1969) 184]; G. Steigman, D. N. Schramm and J. E. Gunn, *Cosmological limits to the number of massive leptons*, Phys. Lett. **66B** (1977) 202.

CHAPTER 15

Inflation

As we have seen, the Hot Big Bang model explains very successfully the existence and properties of the CMB and the primordial nuclear abundances. However, in this chapter we show that it also contains some severe limitations requiring a very important extension: an initial stage of exponential expansion usually referred to as *inflation*.[1]

15.1 THE PROBLEMS OF THE OLD HOT BIG BANG MODEL

There are three big problems within the Hot Big Bang model discussed so far:

- the flatness problem;

- the horizon problem;

- the origin of perturbations.

A fourth problem is the so-called *monopole problem*. It is of historical importance but is not as general as the previous three (it emerges usually within the context of grand-unified theories) and we will not discuss it. We will discuss in this chapter the first two problems, commenting at the end about the role played by inflation for the explanation of the origin of perturbations.

15.1.1 The flatness problem

From the Friedmann equation written in the form (6.45), we can easily derive an equation for the *deviation from flatness* in terms of the energy curvature parameter,

$$|1 - \Omega(t)| = \frac{c^2 \, |k|}{a(t)^2 \, R_0^2 \, H(t)^2} \, . \tag{15.1}$$

This can also be recast as

$$|1 - \Omega(t)| = |1 - \Omega_0| \, \frac{H_0^2}{H^2(t) \, a^2(t)} \, . \tag{15.2}$$

---

[1]Whether inflation is the only solution to these issues is a matter of debate. It is certainly the simplest and most attractive solution found so far.

As we discussed in Chapter 12, at the present time the observations, in particular the position of the first peak in the CMB temperature power spectrum, indicate that our observable universe can be described by a flat geometry with very good precision: $|1 - \Omega_0| < \mathcal{O}(0.1)$.

If we go back in time, during the early universe, we discover that the deviation from flatness had to be incredibly tiny. On more quantitative grounds, let us first calculate the deviation from flatness at the matter-Λ equality time. Since in the last period of the accelerated exponential expansion (de Sitter expansion) the Hubble parameter is constant and since, as we have seen, $a_{\text{eq}}^{M\Lambda} \simeq 0.75$, we can say that at $t_{\text{eq}}^{M\Lambda}$ the deviation from flatness was not much different from the present value, thus $|1 - \Omega(t_{\text{eq}}^{M\Lambda})| \sim |1 - \Omega_0|$.

We can now use the matter-dominated regime scale factor behaviour $a(t) \propto t^{2/3}$, implying $H(t) \propto 1/t \propto a^{-3/2}$, to find

$$|1 - \Omega(t_{\text{eq}})| = |1 - \Omega_0| \frac{a_{\text{eq}}}{a_{\text{eq}}^{M\Lambda}} < \mathcal{O}(10^{-4}). \tag{15.3}$$

In order to properly describe the observed primordial nuclear abundances, we have seen that the initial temperature of the universe should be not much lower than $T_{BBN}^{\text{onset}} \sim 10\,\text{MeV}$. At this temperature the scale factor was $a(t_{BBN}^{\text{onset}}) \sim 10^{-11}$.

Using the radiation-dominated regime expansion law $a \propto t^{1/2}$, one has $H(t) \propto 1/t \propto a^{-2}$ and therefore, from Eq. (15.2), one finds $|1 - \Omega| \propto a^2$. In this way we can write

$$|1 - \Omega(t_{BBN})| = |1 - \Omega(t_{\text{eq}})| \left(\frac{a_{BBN}}{a_{\text{eq}}} \right)^2 < \mathcal{O}(10^{-18}). \tag{15.4}$$

Therefore, we can conclude that the initial deviation of the universe from flatness was necessarily incredibly tiny! Of course this can be regarded just as a fine-tuned necessary condition to avoid, in case of a closed universe, a big crunch occurring much earlier than the current age of the universe or, in case of an open universe, a very cold universe that had no time to form stars and, therefore, life.

One could invoke the so-called *anthropic principle*[2] to justify such a small deviation from flatness. As we will see there is, however, a more attractive and simpler

---

[2]The anthropic principle consists in explaining special fine-tuned values of certain physical quantities not as the result of fundamental physics laws but on the basis of an anthropic selection. This is possible if our observable universe is not unique but a special realisation among a huge number of possible ones. The other necessary ingredient is that the considered quantity has to be a dynamical variable and can have different values in each universe and in fact, when the whole set is considered, it spans all possible values with a certain distribution of probability. Our specific observable universe would then correspond to a very unlikely value but with the virtue to be compatible with the existence of the observer measuring it. Our measurement is then biased by the fact that we necessarily have to live in a special universe allowing life. The anthropic principle clearly implies non-trivial conditions on fundamental physics laws. They have to be such to allow the existence of a huge number of universes (multiverse model) where the physical quantity can take all possible values, even those most unlikely ones allowing the existence of observers. The difficulty to realise such multiverse scenarios has usually discouraged the application of the anthropic principle but there have been various attempts that have gained great consideration.

explanation that, moreover, is also able to solve another problem of the old Hot Big Bang model: the horizon problem.

15.1.2 The horizon problem

In Chapter 10, in Eq. (10.4), we defined $d_{H,0}$, the horizon distance at the present time that corresponds to the current size of the observable universe. At t_0 there is no difference whether this is expressed in comoving or physical lengths, but if we now calculate the horizon distance at some past time t, then we have to distinguish between the two options. If in Eq. (10.4) we replace t_0 with a generic time t, we obtain the *comoving horizon distance*, explicitly

$$d_H^{(0)}(t) \equiv \int_0^t \frac{c\,dt'}{a(t')} \Rightarrow d_H^{(0)}(a) = \int_0^a \frac{c\,da'}{a'^2\,H(a')}. \tag{15.5}$$

Alternatively, the *physical horizon distance* is simply given, using the general relation Eq. (6.1), by

$$d_H(t) \equiv a(t) \int_0^t \frac{c\,dt'}{a(t')} \Rightarrow d_H(a) = a \int_0^a \frac{c\,da'}{a'^2\,H(a')}. \tag{15.6}$$

If we differentiate with respect to time, we obtain

$$\dot{d}_H(t) = c + H(t)\,d_H(t). \tag{15.7}$$

This shows explicitly the two contributions to the speed of the horizon distance variation: the first is simply the speed of light, due to propagation of the message in the comoving system; the second is the velocity of the expansion, that stretches the horizon distance.[3] For a static universe, $H = 0$, one immediately recovers $d_H(t) = ct$. In the case of de Sitter expansion one has $H = $ const, giving $d_H(t) = c\,H^{-1}\left[e^{H\,t} - 1\right]$, showing that for $t \ll H^{-1}$ the contribution from the expansion is negligible but, once $t \gg H^{-1}$, the horizon distance increases at *super-luminal* speed ($\dot{d}_H \gg c$) and indeed $d_H(t) \gg c\,H^{-1}(t)$.

Let us now consider our Lemaitre model, as singled out by the observations. We can equivalently calculate either the comoving or the physical horizon distance. Since in many cases it is convenient to rescale lengths, calculated in the past, at the present time, we opt for the comoving horizon distance to which, for simplicity, we will refer as the horizon distance.

The calculation proceeds very similarly to the calculation of the sound horizon distance shown in Chapter 12 (see Eq. (12.44)). If we first consider the limit $t \to 0$, we can use the radiation-dominated regime, finding

$$d_H^{(0)}(t < t_{\text{eq}}) \simeq \frac{2\,ct}{a(t)} \simeq \frac{c\,H^{-1}(t)}{a(t)} \propto \sqrt{t} \Leftrightarrow d_H^{(0)}(a < a_{\text{eq}}) \simeq \frac{c\,H^{-1}(a)}{a} \simeq \frac{c\,H_0^{-1}}{\sqrt{\Omega_{R,0}}}\,a, \tag{15.8}$$

---

[3]Think of yourself walking on a tapis roulant whose speed increases linearly with the distance from your starting point.

where notice that $c\,H^{-1}/a$ is the *comoving Hubble radius*, a quantity that will play a crucial role in what follows. The expansion velocity is equal to the speed of light for points separated by a distance equal to the Hubble radius $R_H(t) \equiv c\,H^{-1}(t)$. Light is not fast enough to connect two points at a distance above the Hubble radius at a certain time. Therefore, only points that during some period of time were at a distance below the Hubble radius can be causally connected, in agreement with the result highlighted in Chapter 10 after Eq. (10.2). On the other hand, points that at no time were at a distance below the Hubble radius, have never been causally connected.

As we found in Chapter 10, in a radiation universe the comoving horizon distance coincides exactly with the comoving Hubble radius. Since the universe is ever-decelerating, two points that are today at a distance equal to the Hubble radius were necessarily, in the past, at a distance greater than the Hubble radius. This implies that, a region with a given comoving size $\lambda^{(0)}$, becomes causally connected at a time when $\lambda^{(0)} = d_H^{(0)}(t) = R_H^{(0)}$. At this time the scale $\lambda^{(0)}$ enters the horizon becoming *sub-horizon* sized in a way that at any following time $\lambda^{(0)} < R_H^{(0)}$. At earlier times one has $\lambda^{(0)} > R_H^{(0)}$, implying that the region was causally disconnected: The scale $\lambda^{(0)}$ was outside the horizon or *super-horizon* sized. Therefore, the region enters the horizon only once. The picture does not change at times after the matter-radiation equality time, in the matter-dominated regime. In this case the comoving horizon distance evolves with the scale factor as[4]

$$d_H^{(0)}(a > a_{\rm eq}) \simeq \frac{c\,H_0^{-1}}{\sqrt{\Omega_{R,0}}}\,a_{\rm eq} + 2\,\frac{c\,H_0^{-1}}{\sqrt{\Omega_{M,0}}}\,(\sqrt{a} - \sqrt{a_{\rm eq}}) \simeq 2\,\frac{c\,H^{-1}(a)}{a} - \frac{c\,H^{-1}(a_{\rm eq})}{a_{\rm eq}}\,.$$

(15.9)

For $a \gg a_{\rm eq}$, the comoving horizon distance is twice the comoving Hubble radius. This is still monotonically increasing with the expansion. As in the radiation-dominated regime, given a certain comoving scale $\lambda^{(0)}$, this will enter the horizon only once at a specific time and it will then remain sub-horizon. In particular, our current observable universe starts to be causally connected only at the present time but it was never causally connected in the past.

At first sight this result seems harmless, but it actually poses quite a serious problem. We started imposing the cosmological principle as a working assumption justified by the observations. This allowed us to use a simple FRW metric to describe the expansion. However, if we now try to understand the cosmological principle as the result of some physical process, we encounter a basic obstacle in our flat Lemaitre model starting in a radiation-dominated regime. This is because going back sufficiently in the past, there was always a time when a region with a certain comoving size $\lambda^{(0)}$ was necessarily causally disconnected and, therefore, there could be no exchange of causal signal able to explain isotropy and homogeneity. In the case of our observable universe this time is our present time.

---

[4]We can neglect the Λ-dominated regime, since this started only in the relatively recent past and its inclusion would just give a small correction, without changing the overall picture.

One could object that this is not an issue since the cosmological principle would apply only to regions that are sub-horizon. Therefore, waiting a sufficiently long time, once a region becomes causally connected, the FRW metric would be valid. However, this picture is clearly at odds with the observations.

The most evident way to show this point is based on CMB observations. Even though opposite points on the last scattering surface come in causal contact only at the present time, the average temperature of the CMB is basically the same, independently of the direction of observation, over the whole sky. As we have seen the current angular size of a region that was causally connected at the time of recombination is $\sim 1°$. This implies that relic photons coming from two opposite directions of the sky, i.e., separated by an angular distance of 180 degrees, had origin in regions that were not in causal contact at the time of recombination and, more generally, they were never in causal contact in the past. In fact they are not yet causally connected even at the present time!

This means that one cannot find any physical process that, acting in the past, has made the CMB isotropic as we observe it at the present time.[5] Therefore, in the hot Big Bang model that we have discussed so far, the conclusion is that the *cosmological principle* has to be postulated, meaning that the homogeneity and the isotropy of the universe cannot be understood as the result of some physical process.

15.2 AN ELEGANT SOLUTION: INFLATION

Should we desist from understanding flatness, homogeneity and isotropy of the universe and accept them just as sort of built-in features?

Until the end of seventies, there was no convincing solution to these problems. In 1981 Alan Guth showed [1] that an initial phase of de Sitter expansion occurring before the radiation-dominated regime can actually explain both flatness and the homogeneity and isotropy of the universe without having to assume them as special initial conditions. This de Sitter expansion stage, taking place before the radiation-dominated regime, is called *inflation*.[6]

Let us first of all set some notation and introduce useful quantities. Suppose that an inflationary stage starts at some initial time t_i and ends at some final time t_f. In the interval of time $[t_i, t_f]$ the scale factor can be described by (see Eq. (8.15)),

$$a(t) = a_i \, e^{H_i \, (t-t_i)} \qquad (t_i < t < t_f) , \qquad (15.10)$$

where $a_i \equiv a(t_i)$ and $H_i \equiv H(t_i)$. Let us keep in mind that the expansion rate during the de Sitter expansion is constant ($H(t) = H_i$).

---

[5]Notice that it is not only the average temperature to be the same, but even the acoustic peaks are the same if extracted from CMB anisotropies observed in opposite regions of the sky.

[6]A model of inflation, today referred to as *Starobinsky inflation* [2] from the name of its proposer, was presented in a paper submitted a few months before the paper by Guth. Interestingly, the current constraints from CMB temperature anisotropies on the scalar spectral index n_s and on the tensor-to-scalar ratio r seem to support this model. Other important models of inflation are *new inflation* [3] and *chaotic inflation* [4].

It is customary to introduce the so-called *number of e-folds* N in a way that

$$\frac{a_{\rm f}}{a_{\rm i}} = e^{H_{\rm i}\,(t_{\rm f}-t_{\rm i})} \equiv e^N \quad \Leftrightarrow \quad N \equiv \ln\left(\frac{a_{\rm f}}{a_{\rm i}}\right) = H_{\rm i}\,(t_{\rm f}-t_{\rm i})\,. \tag{15.11}$$

Let us see how this assumption is able to explain both flatness and homogeneity and isotropy, without having to assume them as special initial conditions.

15.2.1 Inflation solves the flatness problem

Let us first sketch a qualitative geometric picture. If one starts with a very complicated universe geometry, highly inhomogeneous and anisotropic, a de Sitter expansion acts like a sort of magnification lens able to bring, in a very short time, a tiny microscopic portion of the universe to a macroscopic scale. No matter how harsh the initial geometry is, in this way one can always obtain, at the end of such a rapid period of de Sitter expansion, a homogeneous flat region on macroscopic scales. It is simple to show that the occurrence of such an inflationary stage can solve the flatness problem. If we go back to Eq. (15.2), expressing the deviation of the universe from flatness during the inflationary stage, this gets specialised as

$$|1 - \Omega(t)| = |1 - \Omega_{\rm i}|\,e^{-2\,H_i\,(t-t_i)}\,. \tag{15.12}$$

Therefore, the value of the curvature at the end of the inflation is given by

$$|1 - \Omega_{\rm f}| = e^{-2N}\,|1 - \Omega_{\rm i}|\,. \tag{15.13}$$

If N is sufficiently large, one can always obtain a final value of $|1 - \Omega_{\rm f}|$ that is sufficiently small to be in agreement with all the observations that we discussed, independently of the initial value $|1 - \Omega_{\rm i}|$.

15.2.2 Inflation solves the horizon problem

Let us now calculate the horizon distance during the inflationary stage, neglecting the pre-inflationary stage. For a de Sitter expansion taking place in the time interval $t \in [t_i, t_f]$, one easily finds

$$d_H^{(0)}(t_f) = \frac{c\,H_i^{-1}}{a_i}\,(1 - e^{-N}) \simeq \frac{c\,H_i^{-1}}{a_i} = e^N\,\frac{c\,H_i^{-1}}{a_{\rm f}}\,. \tag{15.14}$$

Therefore, if $N \gg 1$, one has that the comoving horizon distance at the end of the inflation is huge compared to the comoving Hubble radius. If we also add the contribution from the expansion during the radiation-dominated regime, between $t_{\rm f}$ and $t_{\rm eq}$ and from the matter-dominated regime, between $t_{\rm eq}$ and t_0, we obtain for the horizon distance at present

$$d_{H,0} \simeq \frac{c\,H_i^{-1}}{a_i} + \frac{c\,H_0^{-1}}{\sqrt{\Omega_{M,0}}}\,\sqrt{a_{\rm eq}} + 2\,\frac{c\,H_0^{-1}}{\sqrt{\Omega_{M,0}}}\,. \tag{15.15}$$

Considering that in the matter-dominated regime $H \propto a^{-3/2}$, and that in the radiation-dominated regime $H \propto a^{-2}$, we can express H_i^{-1} in terms of H_0^{-1}, obtaining

$$H_i^{-1} = H_0^{-1} a_{eq}^{3/2} \left(\frac{a_f}{a_{eq}} \right)^2 . \tag{15.16}$$

Moreover, neglecting the small contribution from the radiation-dominated phase, we finally find

$$d_{H,0} \simeq 2 \, c \, H_0^{-1} \left(1 + \frac{a_f \, e^N}{2 \, \sqrt{a_{eq}}} \right) . \tag{15.17}$$

For N sufficiently large, the contribution from the inflationary stage dominates, in a way that the size of the causally connected region at the present time is much bigger than our observable universe. In this way the paradox of the CMB properties equal over the entire sky can be solved. Within this picture, our observable universe before inflation was corresponding to a microscopic causally connected region of the universe. During inflation this tiny connected region experienced an exponential expansion reaching a macroscopic size at the end of inflation. At that point the expansion continued in the ordinary radiation-dominated regime. The size of the universe that we observe, basically the distance of CMB last scattering surface, is just a tiny fraction of the entire causally connected region.

It is important also to highlight that during inflation the accelerated expansion brings our observable universe outside the horizon. From that moment the region is causally disconnected and all physical properties acquired in the previous phase, when it was causally connected, remain imprinted until the present time, when our observable universe is re-entering the horizon. This is possible because during the inflationary stage approximately the comoving Hubble radius $R_H^{(0)} \equiv c \, H^{-1}/a \propto a^{-1}$: it decreases with the expansion. If the number of e-folds is high enough, there was a past time before which our observable universe was inside the horizon and causally connected.[7]

This situation is depicted in Fig. 15.1. The comoving Hubble radius $R_H^{(0)}$ (thick solid line) decreases during the inflationary stage in a way that the scale of our observable universe, that became super-horizon during some generic pre-inflationary stage,[8] first exits the horizon and then re-enters it at present. For definiteness we assumed that at the end of inflation the universe enters a radiation-dominated regime straight away but, in general, the transition can be more involved. For a comparison, we also show the evolution of $R_H^{(0)}$ in a universe with no inflation (thin

---

[7]There is an easy life analogy. Consider the case of two childhood friends influencing each other in some early period of their life, then losing contact and memory of each other. One day, many years later, they find some old lost pictures of themselves together hidden in a dusty trunk. This suddenly brings memories of forgotten events that are able to explain still existing traces of that far past.

[8]We have also indicated a pre-inflationary stage, for definiteness in a radiation-dominated regime, even though it should be clear that there is no observational information on this stage. We could also have assumed simply no inflationary stage setting $a_{in} = a_i$ with a given value H_i such to match the Hubble constant (see Eq. (15.16)).

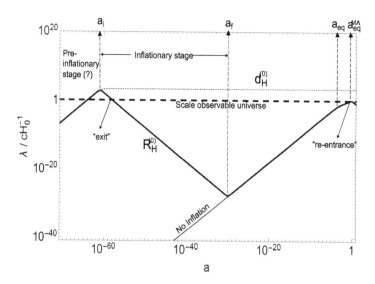

FIGURE 15.1: Evolution of the comoving Hubble radius $R_H^{(0)}$ with the scale factor. During the inflationary stage, in the interval $[a_i, a_f]$, the comoving Hubble radius decreases approximately as $\propto 1/a$. As an effect of the inflationary stage, the length scale corresponding to our observable universe first exits the horizon and then it re-enters it at present. The horizon distance $d_{H,0}$ can be today many orders of magnitude bigger than the scale of our observable universe.

solid line). In this case the comoving Hubble radius would increase monotonically leading to the horizon problem. The figure clearly shows how the inflationary stage solves the horizon problem if the number of e-folds $N \equiv \ln(a_f/a_i)$ is sufficiently large. Notice that we also show the late Λ-dominated regime, showing how the comoving Hubble radius starts again to decrease.

15.2.3 On the importance of an accelerated stage

We want to provide here some additional insight that should help to understand better why an early inflationary stage solves both the flatness and horizon problems. In particular we want to show the importance of having an accelerated stage, not necessarily a de Sitter one.

In order to solve the horizon problem, it is sufficient to have a stage when the comoving Hubble radius, $R_H^{(0)} \equiv cH^{-1}/a$, decreases during the expansion. In this way going back in time, if the ratio a_f/a_i is large enough, there could be a past stage when our observable universe was inside the horizon, it then went out and only now is re-entering the horizon. This is confirmed by the general expression for the integrated horizon Eq. (15.5), that we can rewrite in the form

$$d_H^{(0)}(a) = \int_{a_i}^{a} \frac{da'}{a'} \frac{c\,H^{-1}(a')}{a'}, \qquad (15.18)$$

clearly showing that if $R_H^{(0)}(a) \equiv c\,H^{-1}/a \to \infty$ (or to a constant) for $a \to 0$, then $d_H^{(0)}(t)$ diverges. This again shows that if the ratio $a_{\mathrm{f}}/a_{\mathrm{i}}$ is sufficiently large, then $d_{H,0}$ can be much larger than our observable universe.

A decreasing comoving Hubble radius is also the crucial ingredient to solve the flatness problem, since from the general Eq. (15.2) for the deviation from flatness, one can see that this is $\propto (H^{-1}/a)^2$.

It is then easy to understand that, since $H^{-1}/a = 1/\dot{a}$, one needs $\dot{a}$ to increase during the expansion in order for the comoving Hubble radius to decrease: therefore, an accelerated stage is equivalent to having a decreasing comoving Hubble radius. After all this is quite easy to understand: if the universe is accelerating, a comoving coordinate point that is on the edge of the Hubble radius, will be outside the Hubble radius an instant after and in this way the comoving Hubble radius necessarily decreases during an acceleration stage.

We can also generalise the results obtained in the previous section considering a multi-fluid model with generic fluids described by equation of states of the form $p_i = w_i\,\varepsilon_i$, with $w_i = \mathrm{const}$. In this case one obtains quite straightforwardly

$$\frac{c\,H^{-1}(a)}{a} = \frac{c\,H_0^{-1}}{\sqrt{\Omega_{i,0}}}\,a^{\frac{1+3w_i}{2}}. \tag{15.19}$$

When this expression is inserted into Eq. (15.18), one obtains that in the i-th fluid-dominated regime the comoving horizon distance is given, for $w_i \neq -1/3$, by

$$d_H^{(0)}(a) \simeq d_H^{(0)}(a_{\mathrm{eq}}^{(i)}) + \frac{c\,H_0^{-1}}{\sqrt{\Omega_{i,0}}}\frac{2}{1+3w_i}\left(a^{\frac{1+3w_i}{2}} - (a_{\mathrm{eq}}^{(i)})^{\frac{1+3w_i}{2}}\right), \tag{15.20}$$

where we indicated with $a_{\mathrm{eq}}^{(i)}$ the value of the scale factor at the equality time between the previous fluid-dominated regime and the i-th fluid-dominated regime. If $1 + 3w_i > 0$, then one has an ever-decelerating stage and the comoving horizon distance is dominated by the contribution from highest values of a. On the other hand, if $1 + 3w_i < 0$, one has an ever-accelerating stage and the comoving horizon distance is dominated by the contribution from the lowest values of a close to $a_{\mathrm{eq}}^{(i)}$. The lower $a_{\mathrm{eq}}^{(i)}$ is, the higher the horizon distance.

15.2.4 Final remarks

The power of inflation goes beyond the flatness and horizon problems that originally motivated inflation. Inflation can well explain not only why the universe is homogeneous and isotropic on scales larger than $\sim 100\,\mathrm{Mpc}$ but also the existence of the small primordial perturbations that seeded the formation of the observed galactic structure at smaller scales and its features.

In this case an inflationary stage that is close to a de Sitter stage is necessary in order to produce the correct so-called *spectrum of primordial perturbations* able to explain current observations in quite an astonishing way.

For all these reasons inflation is today regarded as a fundamental ingredient of the ΛCDM model. Many attempts to find alternatives to inflation have been made but they simply fail in matching the observations as successfully as inflation.

Therefore, according to inflation, our observed universe is just a minuscule portion of the entire universe whose global features (e.g., the topology of the universe) could remain unknown forever. On the other hand, a direct test of inflation is challenging. One could hope to find some direct relic trace of inflation. For example, inflationary models typically predict the existence of a relic gravitational waves background. This could be tested in a close future by experiments either indirectly, considering that it would leave an imprint on CMB B-mode polarisation anisotropies (see footnote 10 in Chapter 12), or even directly with gravitational waves interferometers such as LIGO and VIRGO.

It should also be said that a detailed description of inflation based on particle physics encounters many obstacles that do not seem to be circumvented at the moment. In other words, we still do not have a description of inflation based on a fundamental theory. It is however strongly believed that such a theory falls in the realm of new physics, i.e., beyond the standard model and/or general relativity. Therefore, like dark matter and like the cosmological constant, inflation represents today an additional cosmological puzzle requiring new physics.

EXERCISES

Exercise 15.1 *Assume that at the end of the inflationary stage, at t_f, the temperature is given by $T_f = 10^{14}$ GeV. Assume moreover that the number of e-folds is given by $N = 50$. What was the physical radius of the current observable Universe at the beginning of inflation, i.e., at t_i, and at the end of inflation, i.e., at t_f?*

Exercise 15.2 *Consider an inflationary stage occurring in the time interval $[t_i, t_f]$, with $t_f - t_i = 10^{-32}$ s and number of e-folds $N = 100$. What is the value of the expansion rate H_i during inflation (in s^{-1})?*

BIBLIOGRAPHY

[1] A. H. Guth, *The inflationary universe: A possible solution to the horizon and flatness problems*, Phys. Rev. D **23** (1981) 347.

[2] A. A. Starobinsky, *A new type of isotropic cosmological models without singularity*, Phys. Lett. **91B** (1980) 99.

[3] A. D. Linde, *A new inflationary universe Scenario: A possible solution of the horizon, flatness, homogeneity, isotropy and primordial monopole problems*, Phys. Lett. **108B** (1982) 389; A. Albrecht and P. J. Steinhardt, *Cosmology for grand unified theories with radiatively induced symmetry breaking*, Phys. Rev. Lett. **48** (1982) 1220.

[4] A. D. Linde, *Chaotic inflation*, Phys. Lett. **129B** (1983) 177.

CHAPTER 16

ΛCDM model and cosmological puzzles

I n our journey through cosmology, we have seen how data support a specific cosmological model: the ΛCDM model. This is able to describe all cosmological observations with a minimal set of cosmological parameters. In the first section of this chapter we revise and summarise the main features of the ΛCDM model discussed in previous chapters. In the second section we discuss how, though the ΛCDM model provides a perfectly consistent explanation of the cosmological observations, it contains ingredients that cannot be explained by the standard model of particle physics and fundamental interactions. This is a clash that results in the existence of four modern cosmological puzzles providing today, together with neutrino masses and mixing, the strongest evidence of new physics.[1]

16.1 ΛCDM MODEL

Let us summarise the main features of the ΛCDM model.

- It is characterised by a flat geometry, corresponding to an energy density parameter $\Omega_0 = 1$, in agreement with the expectations from inflation;

- Fluids with three different equations of state contribute to the energy-budget at present: matter with $\Omega_{M,0} \simeq 0.31$, dark energy with $\Omega_{\Lambda,0} \simeq 0.69$ and a very small contribution from radiation with $\Omega_{R,0} \simeq 0.915 \times 10^{-4}$;

- The matter contribution, in turn, is the sum of two terms, $\Omega_{M,0} = \Omega_{B,0} + \Omega_{DM,0}$, the first one from ordinary baryonic matter, $\Omega_{B,0} \simeq 0.05$, and the second one from (cold) dark matter, $\Omega_{DM,0} \simeq 0.26$, likely due to the existence of a new elementary particle;

- Dark energy is currently very well described by an equation of state parameter

---

[1]Metaphorically one could say that cosmologists have well tidied up their lounge for the party, but they wiped off all the dust into the adjacent particle physicist's bedroom!

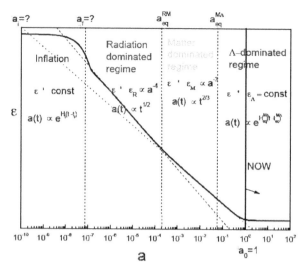

FIGURE 16.1: Evolution of the energy density in the ΛCDM model including the initial inflationary stage. There are four well-distinguished regimes characterised by the dominance of one of the four fluids: inflaton, matter, radiation and the cosmological constant in chronological order. It has to be noticed how the onset of the Λ-dominated regime, characterised by an accelerated expansion, started only very recently: it is sometimes referred to as the *why now problem*.

$w_\Lambda = -1$, identifiable with a cosmological constant in the Einstein equations or vacuum energy density;

- Radiation is made up of photons and of three massless neutrino species; though the contribution at present is negligible, radiation was dominating the early stages of the history of the universe after the end of the inflationary stage and before matter-radiation decoupling;

- In the ΛCDM model, a thermal radiation-dominated regime is preceded by an inflationary stage very closely realising a de Sitter expansion. This is driven by the vacuum energy of a fluid commonly referred to as *inflaton*, likely corresponding to an unknown quantum scalar field. The inflationary stage nicely explains why the universe is flat and solves the horizon problem. Inflation also provides a way to understand the origin of the primordial perturbations, responsible for the observed CMB temperature anisotropies at recombination and, lately, also for the formation of the large-scale structure that we observe today (galaxies, clusters of galaxies and so forth).

With these features, the evolution of the energy density of the universe and of the scale factor within the ΛCDM model is characterised by four stages shown in Fig. 16.1. It should be emphasised how the ΛCDM model, though so successful so far, should not be considered as the ultimate cosmological model. Indeed despite the fact that the latest *Planck* satellite results [1] have found no compelling reasons to extend the ΛCDM model, there are different reasons to believe that new cosmological observations should require extension of the ΛCDM model with the inclusion of new cosmological parameters.

From neutrino mixing experiments we know that at least two neutrinos have to be massive and non-relativistic at the present time. However, current cosmological observations do not find any sign of the neutrino masses and place a stringent upper bound on their sum, $\sum_i m_{\nu_i} \lesssim 0.17\,\text{eV}$ [2]. This is still compatible with the lower bound coming from neutrino mixing experiments, $\sum_i m_{\nu_i} \gtrsim 0.06\,\text{eV}$. However, in the next years it is expected that cosmological observations will become sensitive enough to be able to test this currently allowed window. If they will finally find a signal of non-vanishing neutrino masses, consistent with neutrino mixing experiments, the set of parameters of the ΛCDM model will have just simply to be augmented by one parameter describing the sum of neutrino masses. However, this would be rather a triumph for the ΛCDM model, since neutrino masses would nicely fit into the overall picture requiring just a minimal modification. On the other hand, if cosmological observations find no signal, in disagreement with neutrino mixing experiments, a more drastic modification of the ΛCDM will be necessary. In both cases, this is certainly one of the most exciting tests of the ΛCDM model in the next years.

Another possible future discovery that would require an extension of the ΛCDM model, is the discovery of a non-vanishing value of the so-called tensor-to-scalar perturbation ratio $r \equiv T/S$ from the observation of CMB anisotropies. The data from the *Planck* satellite, combined with those of the Keck and BICEP2 telescopes, have placed a new stringent upper bound $r \lesssim 0.12\,(95\%\,\text{C.L.})$ [8]. However, if future data finally measure a non-vanishing value, this would imply the existence of an inflationary very high energy scale $\sim 10^{16}\,\text{GeV}$, intriguingly close to the grand-unified scale. This parameter would be an indirect trace, imprinted on CMB anisotropies, of the production of primordial gravitational waves during inflation. We might be even so lucky to live to see a new revolution in cosmology, discovering directly primordial gravitational waves. These would allow us finally to test directly an era when the universe was much younger than the earliest stage we access now with Big Bang nucleosynthesis (not much earlier than $\sim 1\,\text{s}$). This would open literally a completely new era in cosmology. For example, we might detect primordial gravitational waves produced during the inflationary stage shedding light on the fundamental physics responsible for inflation. Also we might discover gravitational waves from phase transitions in the early universe which might be connected to some symmetry breaking and, therefore, associated to an extension of the standard model that we know is needed in order to solve the cosmological puzzles, that we are going to discuss in the next section, and also neutrino masses and mixing.

The discovery of gravitational waves made by the LIGO interferometer [3] has therefore opened a new kind of astronomy, which in the future it might prove to be the crucial step toward a fundamental discovery marking the beginning of a new era in cosmology.

Finally another possible, important and certainly desirable, future development beyond the ΛCDM model, is that cosmological observations could provide some information on the nature of dark energy showing some deviations from a pure cosmological constant description. There is a long list of models of dark energy but so far a mere description in terms of a cosmological constant resisted all attempts to falsify it.

16.2 COSMOLOGICAL PUZZLES

Independently of the results of all the experimental tests of the ΛCDM model we discussed, there are very strong conceptual motivations to believe that the ΛCDM model is not the ultimate cosmological model: these are the so-called *cosmological puzzles*. They arise when one tries to combine consistently the ΛCDM model with the standard model and general relativity, discovering that there are unavoidable clashes.

Dark matter is a robust and necessary ingredient in the ΛCDM model; its fundamental nature cannot be explained within established fundamental physics and for this reason it is considered a strong motivation for the existence of new physics. We will discuss in detail, in the next chapter, the most attractive proposals for the solution of the most long-standing of the cosmological puzzles.

Inflation. We have already mentioned, in Chapter 15, how a model of inflation also requires new physics and it, therefore, represents another cosmological puzzle. At the moment a strong test of models of inflation is given by the upper bound on the tensor-to-scalar ratio parameter r, that sets the relative strength of tensorial perturbations, i.e., primordial gravitational waves, compared to scalar perturbations. Many models have been ruled out in this way but the variety of viable models is still quite broad. A measurement of r could certainly in the future provide a precious guidance toward a realistic model of inflation and the physics responsible for it.

Matter-antimatter asymmetry of the universe and models of baryogenesis. A third important cosmological puzzle is represented by the fact that our observable universe seems to be made, barring small traces of anti-matter in cosmic rays, only of matter with no traces of anti-matter. As we discussed, this is the result of a small imbalance between matter and anti-matter that is today observed in the form of a baryon asymmetry. This is what has survived after the sequence of particle annihilations occurred in the early universe while temperature was dropping progressively below the masses of the various elementary particles. The last ones were electron-positron annihilations. This small asymmetry should have been generated after or at the end of inflation in a dynamic generation mechanism, so-called *baryogenesis*[4]. The reason is that any asymmetry generated prior to the inflationary stage would be very efficiently wiped out by inflation.

The reason why the matter-antimatter asymmetry of the universe is indeed a puzzle is that in colliders, starting from a matter-antimatter symmetric initial state, we do not observe any generation of such asymmetry: anti-matter is always produced in the same amount as matter. This is not completely true. There is a small asymmetric behaviour between particles and anti-particles observed in the quark sector and first discovered in the decays of K mesons [5]. This can be associated with the presence of so-called *CP violation* in the quark sector encoded in the Cabibbo–Kabayashi–Maskawa matrix describing quark mixing. However, despite numerous attempts to build viable models such as *electroweak baryogenesis* [6], the small *CP* violation observed in the standard model cannot explain the observed matter-antimatter asymmetry of the universe that, for this reason, is regarded as yet another cosmological puzzle requiring new physics for its solution.

Currently, the most attractive explanation is provided by the same extension of the standard model that is required in order to explain neutrino masses and mixing. This relies on the introduction of very heavy right-handed neutrinos[2] with masses between the electroweak scale ($E_{EW} \sim 100\,\mathrm{GeV}$) and the grand-unified scale ($E_{GUT} \sim 10^{16}\,\mathrm{GeV}$). They can naturally decay asymmetrically into leptons and anti-leptons generating a lepton asymmetry. At temperatures above $100\,\mathrm{GeV}$, this would be partially converted into a baryon asymmetry by so-called *sphaleron transitions*, non-perturbative processes related to the properties of electroweak vacuum, that violate both baryon (B) and lepton (L) number but preserve their difference $B - L$. The asymmetry would be then generated at very high temperatures after inflation. This mechanism is known as *leptogenesis* [7]. Intriguingly, the presence of *CP* violation that would be responsible at high energy for the matter-antimatter asymmetry of the universe is expected in general, though not necessarily, also to be accompanied by *CP* violation in low energy neutrino mixing, that would manifest itself as a different oscillation probability for neutrinos and anti-neutrinos. Latest data seem to support the presence of *CP* violation in neutrino mixing and this can be legitimately regarded as an additional experimental result supporting leptogenesis, though certainly not conclusive. In any case the robust neutrino mixing experimental programme, will certainly help in testing leptogenesis.[3]

Dark energy. A fourth cosmological puzzle, the toughest one, is understanding the nature of dark energy, the mysterious fluid driving the current acceleration of the universe. The problem is not that we do not know, in modern physics, a source for a cosmological constant-like fluid, but that all sources we know would generate a value that is in the best case 10^{20} times larger than the observed value and, in the worst, 10^{122} times larger. Unfortunately the only solution to the problem is to

---

[2]We will discuss them in more detail in the next chapter as possible candidates for dark matter.

[3]On the other hand, the additional discovery of a positive signal in neutrinoless double beta decay experiments could legitimately be regarded as a strong positive test, since it would necessarily imply lepton number violation and the Majorana nature of neutrinos. It would be that strong, that one could legitimately start considering leptogenesis as an established model of baryogenesis. The information from low energy neutrino experiments, combined with leptogenesis, would then provide a powerful phenomenological way to single out a realistic model of new physics, maybe even without needing new physics at colliders, especially if some additional evidence of new physics emerges, like for example discovery of primordial gravitational waves or the discovery of a non-astrophysical component in very high energy neutrinos.

assume that there are at least two contributions cancelling each other out with an incredible precision, generating the biggest fine-tuning problem in modern science.

The cosmological puzzles show how cosmology today is an incredibly powerful laboratory to test theories beyond the standard model and general relativity. It is probably the best way we have today to solve fundamental problems in physics related to energy scales too high to be accessible in conventional laboratories, especially considering that no signs of new physics have been found at the LHC so far. In our journey, we have seen how long-standing issues in cosmology have found a clear answer thanks to a very successful combined effort of theoretical ideas and observations. On the other hand, the cosmological puzzles provide a new set of challenging questions to be answered not simply at a cosmological level, but, more generally, at a fundamental one.

BIBLIOGRAPHY

[1] P. A. R. Ade et al. [Planck Collaboration], *Planck 2015 results. XIII. Cosmological parameters*, Astron. Astrophys. **594** (2016) A13 [arXiv:1502.01589].

[2] N. Aghanim et al. [Planck Collaboration], *Planck intermediate results. XLVI. Reduction of large-scale systematic effects in HFI polarization maps and estimation of the reionization optical depth*, Astron. Astrophys. **596** (2016) A107 [arXiv:1605.02985 [astro-ph.CO]].

[3] B. P. Abbott et al. [LIGO Scientific and Virgo Collaborations], *Observation of gravitational waves from a binary black hole merger*, Phys. Rev. Lett. **116** (2016) 6, 061102 [arXiv:1602.03837 [gr-qc]].

[4] A. D. Sakharov, *Violation of CP invariance, c asymmetry, and baryon asymmetry of the universe*, Pisma Zh. Eksp. Teor. Fiz. **5** (1967) 32 [JETP Lett. **5** (1967) 24] [Sov. Phys. Usp. **34** (1991) 392] [Usp. Fiz. Nauk **161** (1991) 61].

[5] J. H. Christenson, J. W. Cronin, V. L. Fitch and R. Turlay, *Evidence for the 2π decay of the K_2^0 meson*, Phys. Rev. Lett. **13** (1964) 138.

[6] V. A. Kuzmin, V. A. Rubakov and M. E. Shaposhnikov, *On the anomalous electroweak baryon number nonconservation in the early universe*, Phys. Lett. **155B** (1985) 36.

[7] M. Fukugita and T. Yanagida, *baryogenesis without grand unification*, Phys. Lett. B **174** (1986) 45.

[8] P. A. R. Ade et al. [BICEP2 and Planck Collaborations], *Joint analysis of BICEP2/Keck array and Planck data*, Phys. Rev. Lett. **114** (2015) 101301 [arXiv:1502.00612 [astro-ph.CO]].

CHAPTER 17

Dark matter

In the last chapters we have seen how, from the knowledge of particle, atomic and nuclear physics, it is possible to describe two fundamental stages in the history of the universe: recombination and nucleosynthesis. However, all observations indicate that there must have been an earlier stage responsible for the origin of dark matter. The ways in which this stage might have proceeded are closely related to the unknown nature of dark matter. Therefore, dark matter reveals a new fascinating aspect of the physics of the early universe: it can be used as a laboratory to test models of new physics that predict the existence of new elementary particle species, including one (maybe even more than one?) that can play the role of dark matter particle. In this chapter we discuss the most attractive models of dark matter, starting from traditional WIMPs, and the strategies that are explored to search for dark matter. Even though unsuccessful so far, they have placed important constraints that guide current theoretical investigation.

17.1 EXISTENCE OF A NON-BARYONIC DARK MATTER COMPONENT

Before the precise and accurate measurements of $\Omega_{B,0}$ and $\Omega_{M,0}$ from CMB anisotropies, it was debated whether a baryonic contribution from small astronomical objects invisible to telescopes, such as brown dwarfs, giant planets, cold stellar remnants, belonging to the category of MACHOs, could account for the whole dark matter.

However, we have seen that, from CMB anisotropies, one finds $\Omega_{B,0} \simeq 0.05$, a result supported also from standard BBN. One also finds, in agreement with galaxy cluster dynamics determination, $\Omega_{M,0} \simeq 0.31$. A comparison clearly indicates that baryonic matter contributes just to $\sim 1/6$ of the total matter energy density of the universe. Therefore, one can conclude that most of the matter is in the form of an unknown form of non-baryonic and non-luminous matter (more precisely neither absorbing nor emitting electromagnetic radiation) commonly referred to as *dark matter*.[1]

---

[1]From a comparison of $\Omega_{B,0}$ with $\Omega_{gas,0}$ and $\Omega_{\star,0}$ one can conclude that there is also some baryonic dark matter contribution, but this only accounts for $\Omega_{B,0} - \Omega_{gas,0} - \Omega_{\star,0} \simeq 0.016$.

It should be stressed that dark matter is needed not only to explain the dynamics of clusters of galaxies and galactic rotation curves, but had to be already present during recombination in order to reproduce the CMB acoustic peaks. In addition, the dark matter component also played a crucial role in galaxy formation, and, for a successful description, it is necessary that it was produced prior to matter-radiation equality time. For these reasons the production of dark matter had to occur in the early universe. We want now to discuss the most popular solutions that have been proposed to solve the dark matter conundrum.

17.2 WHICH SOLUTION TO THE DARK MATTER PUZZLE?

Let us now discuss a few popular solutions that have been proposed to solve the dark matter puzzle.

17.2.1 Modified Newtonian dynamics (MOND)

There have been attempts to explain the galactic rotation curves not as the effect of the presence of a non-baryonic dark matter component, but in terms of a modification of Newtonian dynamics. In 1983 Milgrom [4] proposed that the acceleration due to the gravitational force, Eq. (11.8), should be modified in a way that

$$\frac{f}{m} = \frac{G M}{R^2} = a \, \mu \left(\frac{a}{a_0} \right) , \tag{17.1}$$

where $\mu(a/a_0)$ is a correcting factor such that $\mu(a/a_0) = 1$ for $a \gtrsim a_0 \simeq 10^{-10} \, \mathrm{m\,s^{-2}}$ and $\mu(a/a_0) = a/a_0$ for $a \lesssim a_0$. In this way, for small values of the acceleration, corresponding to large distances, the usual Newtonian dynamics is modified and this can explain the observed galactic rotation curves.

However, this interpretation does not seem to match with other phenomenological evidence for the existence of dark matter. For example, the galactic clusters dynamics seems to be in disagreement with the MOND proposal. In particular a very famous recent discovery provides a striking example. The *bullet cluster* is an astronomical system formed by two colliding clusters of galaxies. The displacement between the hot gas emitting X-rays and the bulk of the dark matter, probed by the gravitational lensing, shows how the matter in clusters of galaxies has to be described in terms of two different fluids exhibiting two quite distinguished behaviours during the collision: ordinary baryonic matter in hot gas dissipates slowing down due to ordinary interactions, while the collisionless dark matter[2] can only interact gravitationally or through very weak forces not playing any role in the bullet cluster. This kind of object cannot be explained within MOND. A more general covariant generalisation of MOND called TeVeS, a tensor-vector-scalar field theory

---

[2]Usual CDM is assumed to be collisionless but recently the possibility to have a self-interacting kind of dark matter is seriously investigated in order to solve anomalies in the properties of satellite galaxies. In this case systems such as the bullet cluster help placing constraints on the strength of self-interactions.

extension of general relativity [5], might survive the bullet cluster probe with the help of massive neutrinos but with an unattractive fine-tuned choice of the parameters. This solution is moreover unable to reproduce CMB acoustic peaks. MOND is, therefore, currently strongly disfavoured (if not ruled out) and a solution to the dark matter puzzle as a genuine new form of matter is still considered much simpler and attractive, despite the difficulties to detect it beyond its gravitational effects.

17.2.2 Primordial black holes

It has been proposed that dark matter might consist of a population of black holes, since they would not contribute to the baryonic component probed by CMB anisotropies and BBN. Dark matter black holes cannot be in any case ordinary astrophysical black holes forming from the collapse of massive stars, but they have to be *primordial black holes*, produced in the early universe, prior to matter-radiation equality time, by some mechanism. The most attractive idea is that primordial black holes would be topological defects leftover from some phase transition that occurred in the early universe.

In this way a primordial abundance of black holes could explain the dark matter contribution $\Omega_{DM,0} \simeq 0.26$. However, primordial black holes would also behave as MACHOs in the galactic halo and gravitational microlensing searches constrain their abundance. The EROS-2 survey places a lower bound of $\sim 15 \, M_\odot$ on the mass of MACHOs able to explain the total dark matter contribution $\Omega_{CDM,0}$. It is not easy for such supermassive black holes to form in an astrophysical way. The discovery by the Laser Interferometer Gravitational-Wave Observatory (LIGO) in 2016 of gravitational waves produced by the merging of two black holes of masses above $\sim 15 \, M_\odot$ [6] has attracted great attention in the community,[3] triggering the question whether this detection should be also regarded as a discovery of dark matter [7]. However, recently it has been noticed that dark matter MACHOs with masses in the range 15–100 $M_\odot$ would disrupt compact stellar systems in ultra-faint dwarf galaxies within cosmological times. The existence of such systems would then close even this allowed window on MACHOs dark matter masses (including primordial black holes) [8]. In this way, when all constraints are combined, MACHOs dark matter masses from $10^{-7} \, M_\odot$ up to arbitrarily high masses are disfavoured.[4]

---

[3]The signal nicely matches the waveform predicted by general relativity for the merger of two black holes with masses $36^{+5}_{-4} \, M_\odot$ and $29^{+4}_{-4} \, M_\odot$ at a distance of 410^{+160}_{-180} Mpc (corresponding to a redshift $z = 0.09^{+0.03}_{-0.04}$). The difference between the final mass and the total initial mass, $\Delta M \, c^2 = 3^{+0.5}_{-0.5} \, M_\odot \, c^2$ gives the energy radiated in gravitational waves.

[4]This of course does not exclude the possibility that LIGO has discovered primordial black holes contributing sub-dominantly to dark matter. Moreover the constraints apply to a monocromatic population of black holes, i.e., with the same mass, but they could be relaxed in the case of a broad distribution of masses.

17.2.3 Massive neutrinos

The most attractive solution, satisfying all requested properties and constraints, is that dark matter is made of some massive elementary particle with no electromagnetic or strong interactions. The big puzzle is to identify its nature determining quantum numbers, mass, charges. It is natural first to scrutinise the zoo of known elementary particles. Among them, only neutrinos have the required properties and they were indeed the first candidate for particle dark matter to be considered in the late seventies. However, they need to be massive, contrarily to the standard model massless neutrinos.

We know that in the standard model, for each of the three charged leptons, there is a corresponding active neutrino forming so-called lepton doublets. Therefore, we have electrons and electron neutrinos, muons and muon neutrinos and finally tauons and tauon neutrinos. How can we calculate the neutrino contribution to $\Omega_{M,0}$?

In the case of massless neutrinos we have seen that if the initial temperature of the universe is higher than $T^\nu_{\text{dec}} \sim 1\,\text{MeV}$, then the three neutrino species would thermalise and after decoupling, if stable, they would survive until the present with an abundance given by $(\alpha = e, \mu, \tau)$

$$n_{\nu_\alpha,0} = \frac{3\zeta(3)}{2\,\pi^2} \frac{T^3_{\nu,0}}{(\hbar c)^3}\,, \tag{17.2}$$

where, as we discussed in Chapter 12, $T_{\nu,0} = (4/11)^{1/3} T_0 \simeq 0.7\,T_0$.

If we now consider neutrinos with masses $m_{\nu_i} c^2 \ll T_{\nu,\text{dec}}$ but still $\gtrsim T_{\text{eq}} \sim 1\,\text{eV}$, their relic number density would still be given by the expression (17.2), as for massless neutrinos. However, now there would be a time when relic neutrinos would become non-relativistic. At this stage neutrinos would stop being a component of radiation and start to contribute to matter, or better, to dark matter, since they are electrically neutral and, if stable, unable to produce electromagnetic radiation. The contribution to the energy density of the universe from massive neutrinos would then be different from the result found in Chapter 12 for massless neutrinos. In general, neutrino mass eigenstates, that we can indicate with ν_1, ν_2 and ν_3 respectively with masses m_{ν_1}, m_{ν_2} and m_{ν_3}, do not necessarily coincide with the flavour eigenstates ν_e, ν_μ and ν_τ but they might mix. In our case we have to focus on the mass eigenstates. Their current number density $n_{\nu_i,0}$ would be exactly given by the same expression Eq. (17.2) calculated for the flavour eigenstates. The energy density in massive neutrinos would then be simply given by $(i = 1, 2, 3)$

$$\Omega_{\nu,0}(m_{\nu_i}) = \frac{\sum_i m_{\nu_i} c^2 n_{\nu_i,0}}{\varepsilon_{c,0}} \simeq \frac{\sum_i m_{\nu_i}}{93.1\,h^2\,\text{eV}/c^2} \simeq \frac{\sum_i m_{\nu_i}}{43\,\text{eV}/c^2}\,. \tag{17.3}$$

This expression is very interesting because it suggests that three active neutrinos with degenerate masses $m_i \simeq 14\,\text{eV}/c^2$ would basically, together with the baryon component, give a flat matter universe with $\Omega_{\nu,0} \simeq 0.95$. However, at the beginning of the 1980s, much earlier than the discovery of dark energy and the rise of the

ΛCDM model with $\Omega_{M,0} \simeq 0.31$, it became clear that such a neutrino-dominated universe could not reproduce the observed large-scale structure of the universe. This is because in a neutrino-dominated universe the large-scale structure would form in a *top-down scenario*, where first big structures form (super clusters of galaxies), and then, by fragmentation, smaller structures originate in different steps (clusters of galaxies, galaxies, etc.). This scenario is clearly disproved by observations showing that protogalaxies, typically in the form of quasars, are the first objects to form[5] followed, by *aggregation*, by clusters and super clusters (*bottom-up scenario*).

The reason why neutrinos would give rise to a wrong bottom-up scenario is that they behave as *hot dark matter*, i.e., they are ultra-relativistic at the matter-radiation equality time. This implies that at that time neutrinos would free-stream, wiping-out all perturbations created during the inflationary stage with comoving scales below ~ 100 Mpc, the scale corresponding to super clusters of galaxies. This negative conclusion on the role of massive neutrinos in structure formation became evident at the beginning of the 1980s, both with analytical methods and with first N-body numerical simulations able to simulate how large-scale structure formed in the universe.

The numerical N-body simulations confirmed in a completely independent way the necessity of the existence of dark matter in order to understand how galaxies could form so quickly after the matter-radiation decoupling. At the same time they showed that in order to reproduce correctly the bottom-up scenario supported by the observations, dark matter had to be *cold*. This means that the free-streaming length of dark matter particles has to be, at the time of matter-radiation equality time, much shorter than the galactic comoving scale (before getting bound).[6]

In the mid 1990s, one of the most attractive scenarios for structure formation was a flat universe with a mixed cold-hot dark matter component [9]. In this scenario massive neutrinos still played the role of providing a sub-dominant hot dark matter component with $\Omega_{\nu,0} \simeq 0.1$. From Eq. (17.3), one can immediately see that this would correspond to having three degenerate neutrinos with (individual) masses $m_{\nu_i} \simeq 1.5$ eV.

The discovery of neutrino mixing in atmospheric neutrinos in May 1998 and subsequently also in solar neutrinos in 2001, firmly established that neutrinos are massive and more specifically the most recent experimental data from neutrino mixing experiments place a lower bound

$$\sum_i m_{\nu_i} \gtrsim 0.06 \, \text{eV} \, . \tag{17.4}$$

This result, together with the fascinating consequence that the standard model needs to be extended, seemed to provide solid support for a cosmological role of neutrinos as a hot dark matter component, as in the flat mixed hot-cold dark

---

[5]Currently the record for oldest structure ever observed is held by galaxy (GN-z11) with redshift $z \simeq 11$ corresponding to an age of the universe of approximately 350 millions of years.

[6]This scale is about ~ 1 Mpc, a few times bigger than the current typical galactic scale: can you explain why?

matter scenario. Therefore, the discovery of neutrino oscillations seemed finally to establish the role of massive neutrinos as a hot dark matter component, though sub-dominant. However, just a few months later the announcement of the discovery of atmospheric neutrinos, the discovery of the acceleration of the expansion of the universe from observations of SNIa marked the beginning of the rise of the ΛCDM scenario. The flat hot-cold dark matter scenario fell into oblivion and with it also the idea of a significant, though sub-dominant, hot dark matter component.

Within the ΛCDM model a hot dark matter component is not only unnecessary but even unwanted. The reason is that again the presence of an even small quantity of hot dark matter would suppress the formation of structure below 1 Mpc, corresponding basically to dwarf galaxies, in disagreement with the observations. If we define the hot dark matter fraction as

$$f_\nu \equiv \frac{\Omega_{\nu,0}}{\Omega_{DM,0}}, \tag{17.5}$$

today the *Planck* satellite data, when combined with high ℓ CMB anisotropies and baryon acoustic oscillations observations, place a very stringent upper bound (95% C.L.)

$$f_\nu \lesssim 0.02, \tag{17.6}$$

translating, from Eq. (17.3), into an upper bound [2]

$$\sum_i m_{\nu_i} \lesssim 0.17 \, \text{eV}. \tag{17.7}$$

Therefore, even though the result (17.4) from neutrino oscillations places a lower bound $f_\nu \gtrsim 0.006$, the discovery of neutrino masses in neutrino oscillations sounds more like a Pyrrhic victory for a cosmological role of massive neutrinos as hot dark matter.[7]

17.2.4 The WIMP miracle

From what we have discussed so far, we can say that, based on the existing experimental information, currently the most favoured picture is that dark matter is made of a new elementary particle, yet to be detected and identified in laboratory experiments. We will refer to it as the *dark matter particle* and, as we have seen, it has to be sufficiently *cold* to give rise to a bottom-up scenario in structure formation. This conclusion explains the huge interest in dark matter in particle physics, an interest that triggered one of the most fascinating hunts in scientific history: the quest for the nature of dark matter. So far all evidence for dark matter comes from its gravitational effects and these, except for gravitational microlensing, do not tell us

---

[7]There is no reason to be sad for neutrinos. As we discussed in the previous chapter, massive neutrinos might find their revenge with leptogenesis, playing a different, in fact even more fundamental, role in cosmology: explaining the matter-antimatter asymmetry of the universe! In this respect, it is quite encouraging that leptogenesis strongly favours neutrino masses, $m_{\nu_i} \lesssim 0.1 \, \text{eV}$ [2], an upper bound now fully positively tested by the upper bound on the sum of neutrino masses from CMB anisotropies, implying, for each neutrino mass, $m_{\nu_i} \lesssim 0.065 \, \text{eV}$.

about the mass and other properties of the dark matter particles. In particular, we do not know if they have additional interactions beyond the gravitational one. These in any case have to be much weaker than electromagnetic and strong interactions.

Extending the result obtained for the number density of massive neutrinos, Eq. (17.2), we can easily obtain an expression relating the dark matter particle number density to their mass. Using the result found from the fit of CMB acoustic peaks on the abundance of dark matter, $\Omega_{DM,0}\, h^2 \simeq 0.1$, a stable dark matter particle with mass $m_{\rm DM}$ has at present (order-of-magnitude wise) a number density abundance

$$n_{\rm DM,0} \sim 1\,{\rm m}^{-3}\left(\frac{\rm GeV}{M_{\rm DM}}\right) \sim 3\times 10^{-9}\, n_{\gamma,0}\left(\frac{\rm GeV}{M_{\rm DM}}\right). \qquad (17.8)$$

This means that one GeV dark matter particle per cubic meter would reproduce the observed dark matter abundance.[8] Clearly the heavier the dark matter particle is, the lower the needed number density.

Despite the fact that dark matter particles today can be considered, in most cases, as a collisionless gas, they had to be produced in the early universe and this usually requires that, in general, they have to experience some interaction beyond the gravitational one, except for specific production models where the gravitational interaction itself is responsible for their production. The production mechanism has to rely on some interaction of the dark matter particles and within a specific model one can relate the dark matter abundance to the strength of the interaction.

For example we have seen that in the case of light active neutrinos, with masses well below $T^\nu_{\rm dec}$, weak interactions are responsible for the production of the dark matter in pair production processes such as $e^+ + e^- \to \nu_\alpha + \bar{\nu}_\alpha$. This is a thermal production mechanism, since neutrinos are produced from thermal bath interactions. In the case of light neutrinos it was simple to calculate their relic abundance, since it coincides with the thermal abundance at the time of decoupling and, since $T^\nu_{\rm dec} \simeq 1\,{\rm MeV} \gg m_{\nu_i}c^2$, the ultra-relativistic limit applies.

Let us now consider a hypothetical *weakly interacting massive particle* (WIMP) X with mass m_X, such that $m_X c^2 \gg T^\nu_{\rm dec}$. First examples of WIMPs, studied in the mid seventies, were massive active neutrinos. At that time only electrons and muons and their corresponding neutrinos were known but it was quite reasonable to assume that additional lepton families, heavy charged lepton ℓ's with their corresponding neutrinos ν_ℓ's, could exist. There was a particular interest to place constraints on the mass of the ν_ℓ's, calculating their relic abundance $\Omega_{\nu_\ell,0}\, h^2$ as a function of the mass m_{ν_ℓ} and imposing that this was not exceeding the value of the matter abundance. This made it possible to place a lower bound $m_{\nu_\ell} \gtrsim 2\,{\rm GeV}/c^2$ [10, 11, 12]. However, in August 1977, the tauon lepton was discovered and it was possible to place an upper bound on the mass of the tauon neutrino, $m_{\nu_\tau} \lesssim 0.6\,{\rm GeV}/c^2$, ruling out ν_τ as a WIMP. At the same time the LEP 2 measured the number of active neutrino

---

[8]This of course should be meant as an average number density. Dark matter, like baryonic matter, also clumps, for example in galactic halos surrounding visible galaxies. This is an important point on which we will be back soon in more detail when we discuss indirect searches of dark matter.

species with a mass below half of the Z boson mass finding $N_\nu^{LEP} \simeq 3$. This rules out the existence of a fourth heavy active neutrino with a mass below $M_Z/2 \simeq 45\,\mathrm{GeV}/c^2$ and nowadays the LHC also has definitively ruled out a fourth family with an even heavier active neutrino.

However, there is a strong motivation to extend the standard model in a way that it naturally leads to the existence of WIMPs with masses in the range $m_X \sim (10\,\mathrm{GeV}\text{–}1\,\mathrm{TeV})/c^2$: this is the so-called *hierarchy problem* and the related *naturalness* problem. It relies on the observation that in the standard model a value of the electroweak scale $E_{EW} \sim 100\,\mathrm{GeV}$,[9] much lower than the Planck scale, is highly unnatural. This is because it is possible only as a result of a huge fine-tuned cancellation between the *tree-level* Higgs mass and its quantum correction, each about 34 orders of magnitude bigger than the electroweak scale itself.[10] This problem can be solved by introducing appropriate new physics around the electroweak scale. For this reason one expects the existence of new particles with masses not too different from the electroweak scale and one of them, reasonably the lightest, can naturally play the role of WIMP dark matter particle. Let us have then in mind this modern motivation of WIMP dark matter particle and try to understand whether its relic abundance can indeed reproduce the observed dark matter abundance. We need to discuss, qualitatively, how to calculate the relic abundance of WIMPs.

First of all assume a so-called *thermal production* of WIMPs, requiring an initial temperature of the early universe comparable or higher than the mass of the WIMP particle in the range $m_X c^2 \sim 10\,\mathrm{GeV}\text{–}1\,\mathrm{TeV}$. At such high temperatures the abundance of WIMPs is given by the usual ultra-relativistic thermal equilibrium abundance. When T drops below $m_X c^2$, thermal equilibrium implies that its initial ultra-relativistic abundance starts to be Boltzmann suppressed, as $\propto \exp[-m_x c^2/T]$. If one naively assumes that the freeze-out temperature of the WIMP abundance is approximately given by the decoupling temperature of light neutrinos, $T_f^X \simeq T_{dec}^\nu$, then the relic abundance of the X's would be simply given by the abundance of light neutrinos times a Boltzmann suppressing factor $\exp[-m_x c^2/T_{dec}^\nu]$. For particles with a $\sim 100\,\mathrm{GeV}$ mass this would result unavoidably into a completely negligible value of $\Omega_X h^2 \lll \Omega_{DM} h^2$.

However, the assumption $T_f^X \simeq T_{dec}^\nu$ is incorrect, since T_{dec}^ν was calculated assuming that light neutrinos decouple when they are fully ultra-relativistic, and since $T_{dec}^\nu \ll m_X c^2$, this assumption is not valid. In fact the abundance of WIMPs, similarly to what we have seen in BBN for the neutron-to-proton ratio, *freezes out* when they are non-relativistic at a *freeze-out temperature* $T_f^X \simeq m_X c^2/x_f$, with $x_f \simeq 25$, much higher than T_{dec}^ν (i.e., much earlier).

The reason is that the relic abundance is determined by the annihilations of WIMP particles with their anti-particles, assuming that there is no initial asymmetry between their abundances. In this case, when the temperature drops below

---

[9]Order-of-magnitude-wise it is given by the Higgs vacuum expectation value.

[10]The quantum correction is generated by loop diagrams and in the standard model this is proportional to the square of the cut-off scale on the momentum of particles in the loops. The cut-off scale is given by the Planck mass itself and its presence originates the huge quantum correction.

their mass, the WIMP abundance becomes so small that the rate of annihilations, proportional to the square of the particle number density, drops more rapidly than in the case of ultra-relativistic light neutrinos. In first approximation, the value of x_f can be derived using a simple instantaneous decoupling approximation, imposing the usual criterion $\Gamma_X(T_f^X) \simeq H(T_f^X)$, but taking into account the non-relativistic Boltzmann suppression factor for their number density [10].

A more sophisticated calculation from kinetic theory, able to describe statistically how a system evolves out-of-equilibrium, gives a more accurate result. The simplest and most traditional tool for kinetic calculations is provided by the famous Boltzmann equations. Under the given assumptions and some approximations, it can be shown that these reduce to a rate equation for the number density of WIMPs $n_X(t)$, the Lee–Weinberg equation, given by [11]

$$\frac{dn_X}{dt} + 3\,H\,n_X(t) = -\langle \sigma_{\mathrm{ann}}\,v_{\mathrm{rel}} \rangle \left[n_X^2(t) - (n_X^{\mathrm{eq}})^2(t) \right], \qquad (17.9)$$

where $\langle \sigma_{\mathrm{ann}}\,v_{\mathrm{rel}} \rangle$ is the value of the thermally averaged annihilation cross section times the Möller velocity, a relativistic definition of relative velocity between two particles in a generic reference frame. From an analytic solution of this equation, one finds that the final contribution to the energy density parameter from the WIMP relic abundance is given by [13]

$$\Omega_X\,h^2 \simeq \frac{4 \times 10^{-10}}{\langle \sigma_{\mathrm{ann}}\,\beta_{\mathrm{rel}} \rangle} \left(\frac{\hbar c}{\mathrm{GeV}} \right)^2 \simeq \frac{1.6 \times 10^{-37}\,\mathrm{cm}^2}{\langle \sigma_{\mathrm{ann}}\,\beta_{\mathrm{rel}} \rangle_f}, \qquad (17.10)$$

where $\beta_{\mathrm{rel}} = v_{\mathrm{rel}}/c$. Typical weak values of thermally averaged cross sections are given (order-of-magnitude-wise) by

$$\langle \sigma_{\mathrm{ann}}\,\beta_{\mathrm{rel}} \rangle_f^W \sim 0.1\,\alpha_W^2 \left(\frac{\hbar c}{m_X c^2} \right)^2 \sim 10^{-9} \left(\frac{\hbar c}{\mathrm{GeV}} \right)^2 \left(\frac{100\,\mathrm{GeV}}{m_X\,c^2} \right)^2 \qquad (17.11)$$

$$\sim 4 \times 10^{-37}\,\mathrm{cm}^2 \left(\frac{100\,\mathrm{GeV}}{m_X\,c^2} \right)^2,$$

where $\alpha_W \sim 0.03$ is the dimensionless weak coupling constant at energies $\sim 100\,\mathrm{GeV}$ and where typical relative velocity (in units of c) at the freeze-out is given by $\beta_{\mathrm{rel}} \simeq 0.1$. It is then intriguing that for WIMP masses $m_X \sim (100\,\mathrm{GeV}\text{--}1\,\mathrm{TeV})/c^2$, the expected values from naturalness considerations, one obtains $\Omega_X h^2 \sim 0.1$, nicely reproducing the observed value of the dark matter abundance. Because of this tantalising *WIMP miracle*, for long time WIMPs have been regarded as the most attractive dark matter candidate, also considering the additional feature to be potentially detectable by virtue of their weak interactions.

Before discussing the current strategies to detect WIMPs, let us first briefly mention some of the proposed most popular realistic examples of WIMPs [14, 15], each associated to a specific way to solve the naturalness problem. Supersymmetry is certainly the most popular solution, perhaps also the most elegant one. In supersymmetry each standard model particle has a supersymmetric partner with opposite statistics. This means that if a particle is a boson in the standard model, then its supersymmetric partner is a fermion and vice versa. It can be shown that this generates an additional contribution to the Higgs mass quantum correction cancelling out exactly the contribution from standard model particles. For this reason, the most popular dark matter WIMP candidate is the lightest supersymmetric neutral spin 1/2 fermion, the so-called *neutralino*. This can be made stable by adding a discrete symmetry called R-parity with an associate conserved quantum number with only two discrete values: +1 (even particles) or -1 (odd particles). Supersymmetric and standard model particles have opposite R-parity, minus and plus, respectively. Therefore, a supersymmetric particle cannot decay just into standard model particles and this implies that the lightest supersymmetric particle has to be stable for the same reason protons are stable: namely baryon number is conserved and protons are the lightest baryons. Neutralino as dark matter WIMP was proposed in the eighties [16, 17] and it became such an attractive solution that its discovery was considered just a matter of time. Unfortunately, as we will discuss, intense searches of different kinds have not fulfilled expectations and this has stimulated new ideas for models solving the naturalness problem with their own associated dark matter WIMP candidate.

A popular class of theories providing a solution to the naturalness problem, alternative to supersymmetry, are theories with extra-dimensions. They also usually contain candidates for (thermally produced) dark matter WIMPs, the most popular one being the lightest Kaluza–Klein particle. This is associated to some standard model particle, typically the hypercharged gauge boson B. However, other possibilities have been proposed with fancy names such as *branons* and excited states in warped extra dimensions. Other recent examples of models addressing the naturalness problem, with their associated dark matter WIMPs in brackets, are: i) little Higgs models (T-odd particles); ii) two Higgs doublet models (neutral Higgs boson); iii) twin Higgs models (twin neutral Higgs); iv) dark sector models (dark lightest particle).

17.2.5 WIMP searches

The evidence found so far for the existence of dark matter is uniquely based only on its gravitational effects. The attractive feature of dark matter WIMPs is that they also interact weakly and this implies specific experimental signatures that allow to test the WIMP paradigm. There are three strategies pursued to detect dark matter WIMPs through their weak interactions [15, 18]. If we indicate with SM some generic standard model particle, and with X our WIMP,

(i) In *direct searches* one looks for signals of interactions of the kind

$$X + SM \rightarrow X + SM \; ; \tag{17.12}$$

(ii) In *indirect searches* one looks for signals of interactions of the kind

$$X + X \rightarrow SM + SM \; ; \tag{17.13}$$

(iii) In *collider searches* one looks for signals of interactions of the kind

$$SM + SM \rightarrow X's + SM's. \tag{17.14}$$

Let us briefly discuss how these searches are performed and the latest experimental constraints that have been placed on WIMP parameters (mainly on the WIMP mass m_X and on the interaction cross sections).

17.2.5.1 Direct searches

Dark matter WIMPs, though very rarely, can collide with ordinary matter through their weak interactions. In particular, they can hit nuclei in a detector and transfer to them energy (nuclear recoil energy) that can be detected.[11] Since the WIMP mass is $m_X \sim 100\,\mathrm{GeV}/c^2$ and their velocity $\beta_X \sim 10^{-3}$,[12] the typical transferred energy in the collision is $E \sim m_X c^2 \beta_X^2 \sim 100\,\mathrm{keV}$. These processes, if there, would be very rare. It is literally like finding a needle in a haystack. This is why in order to get more and more stringent experimental constraints, there have been tremendous efforts to increase the size of detectors since this obviously enhances the chances to capture a WIMP crossing the detector. An additional difficulty is that neutrons produced from radioactive elements can mimic dark matter WIMPs since they are neutral and heavy as well. Therefore, it is necessary to place huge detectors in underground laboratories in order to screen them from cosmic radiation, moreover in sites with as low natural radioactivity as possible. In the last three decades, different experiments have been placing more and more stringent limits on the mass and on the cross section with nuclei of the dark matter WIMPs. However, there are models of WIMPs with spin dependent cross section and in this case constraints are weaker. These constraints also depend on the evaluation of astrophysical quantities such as the local density and the velocity distribution of WIMPs at our position in the Milky Way, clearly affected by some theoretical uncertainties.[13] Currently the most stringent constraint comes from the Large Underground Xenon (LUX) experiment.[14] Actually three experiments (DAMA-LIBRA, CRESST-II and CoGeNT) have even claimed positive signals but these have been either later recognised as background events or ruled out by the LUX results, except for marginal regions at low values of the WIMP mass (below $6\,\mathrm{GeV}/c^2$), where sensitivity of current detectors lies. Recently the CDMSlite-II experiment has also ruled out these marginal

---

[11]The Coulomb field of the charged nuclei can convert the nuclear recoil energy into thermal motion (phonons), ionisation or scintillation photons.

[12]This should not be confused with β_{rel}, the relative velocity at freeze-out appearing in Eq. (17.10). In this case β_X is the velocity of dark matter WIMPs in the laboratory reference frame. In a perfectly homogeneous and isotropic universe, dark matter WIMPs should also end up at rest in the comoving reference frame. In this case this would just be the speed of the Earth in the comoving reference frame. However, dark matter particles also gain peculiar velocities in the galactic halo due to the inhomogeneous distribution.

[13]The local energy density of WIMPs at Earth's location in the Galaxy is approximately $\varepsilon_X \simeq 0.3\,\mathrm{GeV\,cm^{-3}}$ and the velocity distribution is assumed to be a Maxwellian distribution.

[14]It is located 1510m underground in the Homestake Mine in Lead, South Dakota.

regions placing an upper bound on spin-independent cross section of WIMPs for values of the mass down to $1 \, \text{GeV}/c^2$.

There are many experimental proposals to improve the sensitivity testing values of the mass below $1 \, \text{GeV}/c^2$, opening the exploration of a new mass range. Also in the typical range of masses between $100 \, \text{GeV}/c^2$ and $1 \, \text{TeV}/c^2$ further bigger detectors are planned with masses in the ton range: the XENON1T experiment (that will be followed by its upgrade XENONnT experiment), the LZ experiment (successor of the LUX experiment), the PandaX-4t experiment and ultimately the DARWIN experiment aiming at a 40-ton target mass. Bigger detectors will have to face the problem of how to be able to disentangle the dark matter signal from the so-called *neutrino floor*, the (currently) irreducible background from solar and atmospheric neutrinos. In this respect there are efforts to identify additional features in the nuclear recoil energy distribution that would allow a *directional detection*, so that for example neutrinos from the sun could be distinguished from dark matter signals.

17.2.5.2 Indirect searches

Instead of searching for interactions of dark matter WIMPs directly, inside the underground detectors, a complementary strategy is to look for the products of dark matter interactions within astronomical environments. The interactions with ordinary matter are in this case unhelpful. In a perfectly homogeneous universe, dark matter annihilations, though not completely turned off, would be so strongly suppressed that it would be impossible to detect any observable effect. However, dark matter, like ordinary matter, clumps on scales smaller than $100 \, \text{Mpc}$. Since the rate of annihilations $\propto \rho_{DM}^2$, one expects that the flux of annihilation products is greatly enhanced in very dense dark matter regions, such as the galactic centre or dwarf galaxies,[15] where the density of dark matter particles is expected to be order of magnitudes above the average value.

First of all one can hope to detect γ-rays (photons with energies equal or just below the WIMP mass rest energy) coming from such dense dark matter regions. These have the advantage that would travel from the production site to us in a straight line, retaining information on the source position in the sky. Therefore, one can target dense dark matter regions to have better chances to distinguish a signal from the background. Ideally one would like to observe a monochromatic line but unfortunately the rate of WIMPs direct annihilations into photons is too weak and what one can realistically detect are photons produced as secondary particles from the primary particles produced in the annihilations. These generate an excess spread on a wide range of energies. In the mass range $(0.1\text{--}1) \, \text{TeV}/c^2$ the most stringent upper bound on the WIMP annihilation cross section has been placed by

---

[15]The galactic centre also overlaps with the centre of the dark matter halo embedding our galaxy, while dwarf galaxies are hosted inside sub-halos of dark matter contained and moving inside the much bigger galactic halo and, as in a kind of Chinese box set, even the existence of smaller dark matter clumps at different scales are predicted by numerical N-body simulations.

the Fermi-LAT space telescope and it basically excludes the typical values expected for thermally produced WIMPs (basically the value at freeze-out in Eq. (17.11). At values of the mass above TeV/c^2, most stringent constraints are placed by the HESS ground based γ-ray telescope but they still allow values $\langle\sigma_{ann}\beta\rangle_f$ as large as 10^{-35} cm^2. The CTA experiment is expected to lower this upper bound of about one order of magnitude in a few years. On the other hand at values $m_X \lesssim 0.1\,TeV/c^2$ the most stringent limits come from the *Planck* satellite observations of CMB anisotropies constraining $\langle\sigma_{ann}\beta_{rel}\rangle$ at the time of recombination. These constraints strongly exclude weak cross sections for thermally produced WIMPs.

A few anomalies have been also reported, in particular the *galactic centre excess* in the Fermi-LAT data compatible with WIMP masses $m_X \simeq (10\text{--}40)\,GeV/c^2$ and $\langle\sigma_{ann}\beta_{rel}\rangle \sim 10^{-36}$ cm^2. However, the signal has not been observed in dwarf galaxies and it has been shown that the excess can also be explained by a population of unresolved millisecond pulsars in the galactic centre. Another anomaly is given by the γ-ray line at 511 keV observed by the INTEGRAL space telescope but in this case astrophysical interpretations are now favoured.

Alternatively, one can search for dark matter products in *charged cosmic rays*. However, in this case particles are deflected by the galactic magnetic fields and would travel to us along a random curvy path inside the galaxy, so that directionality is lost. One can simply compare the measured spectrum to the predicted one within the standard model looking for some excess that would be the result of WIMPs annihilations. Typically, instruments detect electrons and protons, that can originate from many astrophysical sources, and also positrons and anti-protons, that is even more interesting for dark matter searches since these would be produced in equal amounts in dark matter annihilations. Given a source for cosmic rays with a specified energy spectrum, the so-called *prompt* spectrum of particles, the predicted energy spectrum on the Earth is the result of many complicated processes, mainly the diffusion due to the galactic magnetic fields. This is usually calculated solving a *Fokker–Planck equation*, that is a generalised diffusion equation.

Cosmic rays are detected with balloon-type, ground based telescope arrays or satellite-based detectors. In 2009 the PAMELA satellite detector reported an excess in the positron spectrum in the energy range $(20\text{--}200)\,GeV/c^2$. The Fermi-LAT and the AMS-02 satellite detectors confirmed the excess though to a lower level, A dark matter interpretation seems plausible but it encounters great difficulties since the value of the cross section required to explain the excess is much larger than the typical values expected for thermally produced WIMPs (see Eq. (17.11)). Many specific models have been proposed but they are in tension with the constraints from γ-rays so that further experiments are needed for a conclusive verdict.

Neutrinos can also be very useful for indirect searches of WIMPs. Thanks to their weak interactions, WIMPs would dissipate energy scattering off nuclei in central dense regions of celestial bodies such as stars and planets. If their velocity becomes smaller than the escape velocity, they get trapped accumulating gradually in the central dense regions with a capture rate Γ_{capt}. When density gets higher and higher, the annihilation rate Γ_{ann} becomes sufficiently large to balance the capture rate and equilibrium is reached. WIMPs, such as neutralinos, cannot annihilate directly into neutrinos but their annihilation products, e.g., quarks and gauge bosons, would produce them secondarily. Neutrinos are the only particles able to escape the centre of stars travelling in a straight line and retaining information on their production site. In our case the only object able to produce a detectable neutrino flux would be the Sun or, to a lower level, the same centre of the Earth.

Neutrino telescopes, big neutrino detectors able to track the arrival direction of neutrinos, pointing toward the centre of the Sun or the Earth, should then detect a high energy neutrino flux. This strategy is sensitive both to the scattering off nuclei and annihilation cross section of WIMPs. The derived upper bounds on spin independent and on annihilation cross section are less stringent than those derived from direct and γ-rays searches. However, those on spin dependent cross section, derived by combining data from ANTARES, IceCube and SuperKamiokande neutrino telescopes, are the most stringent ones.

In April 2013 the IceCube collaboration reported the discovery of two very high energy neutrino events [19]. A further analysis has then reported twenty-six additional high energy neutrino events, with energies in the range $\sim (10$–$1000)$ TeV [1]. These neutrinos are too energetic (and too many!) to be explained by atmospheric neutrinos produced by cosmic rays scattering off nuclei in the upper layer of the atmosphere and are compatible so far with a isotropic distribution.[16] There is certainly a component that is associated with the cosmic rays of similar energies and whose origin is typically explained in an astrophysical way; for example active galactic nuclei, blazars and merging galaxies are most plausible sources, though they are not sufficient to reproduce the whole observed flux and this might indicate the presence of an unknown astrophysical contribution. However, there is currently an excess at energies ~ 100 TeV that does not seem compatible with canonical cosmic predictions for a generic astrophysical component, usually based on the so-called *Fermi acceleration mechanism* [20]. This excess might be interpreted in terms of a non-canonical new astrophysical component. However, quite intriguingly, it might also be interpreted in terms either in terms of dark matter annihilations or decays, but the latter seems more favoured. In particular, annihilations do not seem to be compatible with the isotropic distribution of the high energy IceCube neutrinos. In any case usual thermal produced WIMP models do not seem able to reproduce the correct abundance explaining the excess. We will come back to this anomaly when we discuss models of dark matter beyond the thermally produced WIMP paradigm.

17.2.5.3 Collider searches

Dark matter WIMPs could also be produced in colliders and currently the LHC has the right energy to test a wide range of masses. However, even if produced, it is not simple then to identify dark matter WIMPs among the multitude of events. Since WIMPs are stable, at least on cosmological scales, they cannot decay inside the

---

[16]Therefore, they are certainly not coming from the centre of the Sun.

detector into detectable standard model particles. It is true that their production would result in missing energy and momentum, but this is not necessarily associated with dark matter WIMPs, since neutrino production also produces the same effect. It is then not easy to distinguish dark matter from neutrinos. Strategies to distinguish a WIMP from a neutrino are mainly based on how the first would couple differently to standard model particles. An additional difficulty is that, though one can claim WIMP discovery at colliders, one still has to prove that this is a *dark matter* WIMP. Things would be easier if the discovery of WIMPs comes along together with the discovery of additional new particles, pointing to a particular extension of the standard model like for example supersymmetry. Moreover one could in principle reconstruct the mass m_X and $\langle \sigma_{ann} \beta_{rel} \rangle$, checking whether they have the right values to reproduce the correct dark matter abundance: this would be quite a strong smoking gun for dark matter WIMPs! In addition, a signal at colliders should also be confirmed in direct and indirect searches and it is actually fair to say that only a simultaneous convergence of different experimental indications would probably provide a conclusive proof, also establishing unambiguously the nature of the dark matter WIMPs and associated new particles.

Unfortunately, so far, there are no signs of new particles in the LHC and not even of missing energy and momentum that cannot be explained with neutrinos. From LHC results one can then place upper bounds on the dark matter production cross sections and these are perfectly compatible with those from direct and indirect searches. It should be said that a hadron collider, such as the LHC, has the disadvantage that the initial energy of the state has some uncertainty, since when protons scatter with each other, actually they do not scatter as a whole particle but the scattering is usually between some of the partons (quarks and gluons) inside the proton and these have a distribution of energy despite the proton energy being fixed. This is a limitation that can be circumvented in a lepton collider and there are different proposals for future ones, though nothing is yet approved.

We can fairly conclude that current experimental results exclude a canonical dark matter WIMP realising the above-mentioned WIMP miracle and, therefore, they strongly motivate modifications or alternative models.

17.2.6 Beyond the WIMP miracle

The null results from searches of dark matter WIMPs basically rule out the WIMP miracle, at least in its canonical realisation. The main reason is that the upper bounds on the cross sections are so stringent to point to much smaller values than those required by the WIMP miracle in Eq. (17.11). From Eq. (17.10) one would then obtain a dark matter abundance much higher than the observed one.

Many ideas have been proposed as an alternative to a traditional WIMP miracle scenario. These can be classified into two categories:

- Dark matter is still a WIMP or some other very similar new particle, but one or more assumptions behind the WIMP miracles have to be revisited;

- Dark matter is not made of WIMPs and one has to explore alternative models.

Let us briefly discuss specific ideas falling within both categories.

17.2.6.1 Relaxing the minimal assumptions

The WIMP miracle is based on the assumption that the freeze-out occurs in isolation, i.e., that it is not influenced by the existence of possible additional new particle species that could also be WIMPs. Relaxing this assumption, one can obtain the correct abundance of dark matter WIMPs, with much lower values of $\langle \sigma \beta \rangle_f$, taking into account three possible effects [21]. A first important one is given by so-called *co-annihilations*, occurring when the mass of the dark matter WIMP is sufficiently close to the mass of a heavier unstable WIMP. Another possibility is that annihilations occur resonantly, at energies close to the Z or Higgs boson mass, and this also tends to enhance the annihilation rate reducing the final relic abundance compared to the non-resonant case and making it compatible with the observed value. Both effects can be realised within supersymmetric models, singling out particular regions in the space of parameters (the so-called co-annihilations and funnel regions) where the neutralino is the lightest supersymmetric particle and its relic abundance reproduces the observed dark matter abundance.

17.2.6.2 Beyond the WIMP paradigm

The list of proposed models of dark matter particles, beyond the dark matter WIMPs freeze-out mechanism, is impressively long. This certainly shows how the *dark matter puzzle* is one of the greatest scientific problems in modern science. Here we necessarily have to limit ourselves to mention the most popular ideas [15, 22].

- *Hidden dark matter and the WIMPless miracle.* Let us start discussing an idea that is a straightforward generalisation of the WIMP miracle. The WIMP miracle can be somehow revisited considering dark matter particles with interactions such that

$$\langle \sigma_{ann} \beta_{rel} \rangle_f = \frac{\alpha^2}{m_X^2} \frac{m_{WIMP}^2}{\alpha_W^2} \langle \sigma_{ann} \beta \rangle_f^{WIMP}, \qquad (17.15)$$

  where with $\langle \sigma_{ann} \beta \rangle_f^{WIMP}$ we indicated the value given by Eq. (17.11) with $m_X = m_{WIMP}$. For $\alpha_X / m_X = \alpha_W / m_{WIMP}$ one still has $\langle \sigma_{ann} \beta_{rel} \rangle_f = \langle \sigma_{ann} \beta_{rel} \rangle_f^{WIMP}$. This means that even though the dark matter particle is not a WIMP, it has the same annihilation cross section of WIMPs. In this way the observed dark matter abundance is still reproduced through the usual thermal freeze-out mechanism. For example, a dark matter particle species with $m_X \sim$ MeV and $\alpha \sim 10^{-5} \alpha_W$ would still have the correct relic abundance but it would escape the tight experimental constraints we discussed.

- *Feeble, extremely or super-interacting massive particles (FIMPS, EIMPS or SWIMPs).* These dark matter particle candidates have interactions much weaker than weak interactions. Nevertheless, they can be still produced thermally. However, their abundance never reaches the thermal value before freeze-out, rather it directly freezes-in to the correct value starting from an initial vanishing abundance [23]. The relic abundance is this time proportional to the annihilation cross section instead of being inversely proportional. In this way one can obtain the correct abundance for the same values of masses as WIMPs but with much smaller cross sections, thus escaping current experimental constraints. Alternatively, their production can occur non-thermally from late decays of heavier particles. For example, one could have an unstable WIMP that, by virtue of the WIMP miracle has the correct relic abundance Ω_{WIMP}. Even though these unstable WIMPs cannot play the role of dark matter, still their abundance can be transferred to a Super-WIMP (cosmologically stable) particle through decays, in such a way that

$$\Omega_{SWIMP} = \Omega_{WIMP} \frac{m_{SWIMP}}{m_{WIMP}}. \tag{17.16}$$

If m_{SWIMP} is not too much smaller than m_{WIMP}, the WIMP miracle still works also for SWIMPs, with the difference that SWIMPs escape direct and indirect search constraints. However, WIMPs could be still produced if some astrophysical environments and their decays have some testable effects in cosmic rays if the decaying WIMP is charged. Particle colliders may also find evidence for the SWIMP scenario in this case. Finally decays of WIMPs into SWIMPs might have an impact on CMB and BBN in the form for example of extra-radiation, sometimes called *dark radiation* and often parameterised in terms of the number of effective neutrino species. In the case of BBN, for lifetimes between $\sim 1\,\mathrm{s}$ and $\sim 1000\,\mathrm{s}$, the products of decays alter in general the values of the primordial nuclear abundances in an unacceptable way yielding constraints on the parameters of the model. However, for some fine-tuned choice of the parameters, the late WIMP decays could even be beneficial, solving the long-standing *lithium problem* in BBN.

A popular example of SWIMP particle is the *gravitino*, the supersymmetric partner of the *graviton*, with spin 3/2. It can play the role of dark matter when it is the lightest supersymmetric particle, since in this case it can be cosmologically stable. It can be produced both thermally, through the freeze-in mechanism, and non-thermally through the decays of the next-to-lightest supersymmetric particle (typically the neutralino). In order to get the correct abundance, the reheat temperature cannot be above $\sim 10^{10}\,\mathrm{GeV}$. For gravitino masses below $\sim 1\,\mathrm{TeV}$ this upper bound on the reheat temperature, becomes a few orders of magnitude more stringent taking into account the constraints from BBN. This is usually referred to as the *gravitino problem* [24].

- *Asymmetric dark matter.* The freeze-out mechanism assumes a vanishing initial asymmetry between the number of dark matter WIMPs and anti-WIMPs. However, as in the case of ordinary matter, some mechanism might have generated an asymmetry in the dark matter sector prior to the freeze-out. If this asymmetry is sufficiently large, then the relic abundance basically corresponds to the same initial asymmetry. This idea is quite attractive since one can also link the problem of generation of the baryon asymmetry, addressed by *baryogenesis*, to the problem of dark matter, obtaining unified models of baryon and dark matter-genesis. The indirect detection rate from annihilation is of course strongly suppressed compared to the standard case (since basically one is left only with particles or anti-particles today) and in this way the stringent constraints holding in the standard freeze-out scenario are evaded. On the other hand their signals in direct detection searches are usually enhanced and many constraints have been placed excluding different regions in the parameter space. In particular the elegant solution of having the same asymmetry in baryonic matter and dark matter, $|\eta_B| \simeq |\eta_{DM}|$, implying $m_X = (\Omega_{DM}/\Omega_B) \, m_p \simeq 5 \, \text{GeV}/c^2$, where $m_p \simeq 1 \, \text{GeV}/c^2$ is the proton mass, is now ruled out by the stringent LUX constraints.

- *Axions*. Another attractive and very popular dark matter candidate emerges naturally from the so-called *Peccei–Quinn mechanism* for the solution of the strong *CP* violation problem in QCD [3]. This mechanism predicts the existence of a new particle, called an *axion*, that would have the right features to play the role of a dark matter particle. Axion particles would be associated with a pseudoscalar field a with coupling to gluons determined by a coupling constant f_a^{-1}, where f_a has the dimension of energy, determining the decay rate of axions and also the same axion mass given approximately by

$$m_a \sim 10^{-5} \, (\text{eV}/c^2) \left(\frac{10^{12} \, \text{GeV}}{f_a} \right). \tag{17.17}$$

In order for axions to be stable on cosmological scales, one needs $f_a \gtrsim 10^6 \, \text{GeV}$ but actually from astrophysical constraints one obtains a much more stringent lower bound $f_a \gtrsim 10^9 \, \text{GeV}$, corresponding to $m_a \lesssim 10 \, \text{meV}/c^2$. The cosmological production of axions would proceed non-thermally through a *vacuum misalignment* mechanism producing an abundance

$$\Omega_a \sim 0.3 \, \theta^2 \left(\frac{f_a}{10^{12} \, \text{GeV}} \right)^{1.18}, \tag{17.18}$$

where θ is the initial vacuum misalignment angle. The correct dark matter density is then achieved for $m_a \sim 10^{-3} \, \theta^2 \, \text{meV}/c^2 \sim (10^{-6} \text{–} 10^{-4}) \, \text{eV}/c^2$. Despite the small mass, axions would behave similarly to cold dark matter rather than hot dark matter, in agreement with structure formation constraints. This is because dark matter axions would basically form a Bose–Einstein condensate with negligible kinetic energy.

The axion can be detected directly since it interacts with standard model particles. Currently the most stringent constraints on the mass and coupling constant are placed by the conversion of dark matter axions to photons through scatterings off a background magnetic field (Primakoff process).

Within a supersymmetric scenario, the *axino*, the supersymmetric partner of the axion, is also a viable candidate for dark matter and, contrarily to the axion, it would behave as a SWIMP, similarly to the gravitino.

- *Light sterile neutrinos.* Sterile neutrinos are neutral particles that are singlets in the standard model, i.e., they have no electroweak or strong interactions, and can mix with ordinary neutrinos. The most straightforward examples are right-handed neutrinos that are usually introduced to explain neutrino masses and mixing.

  The production of sterile neutrinos can proceed through the mixing with the same active neutrinos, taking into account medium effects in the early universe that can strongly enhance it. This is the (*Dodelson and Widrow mechanism*), working for typical values of the sterile neutrino mass $m_{\nu_s} \sim$ keV and mixing angle $\theta_s \sim 10^{-4}$. In this way the mixing in vacuum is sufficiently small that there is no (excluded) effect on neutrino oscillation experiments. The sterile neutrino can usually decay radiatively into photons. A quite clean signature would then be the discovery of a line in X-rays within astrophysical environments where there are big quantities of dark matter (super clusters of galaxies, galactic halos and sub-halo).

  Intriguingly, in 2014, an excess around ~ 3.5 keV in X-rays coming from different galaxy clusters has been claimed to have been found in different analyses though it has not been confirmed by all observations, for example by the Hitomi satellite. An interpretation in terms of decays of light sterile neutrino dark matter consistent with the observations is possible, but it requires some sophistication of the Dodelson–Widrow mechanism, for example introducing a large lepton asymmetry that would induce a more efficient resonant production of the light sterile neutrinos [25]. The 3.5 keV anomaly, if confirmed, could then be regarded as strong support for a sterile neutrino dark matter scenario but more experimental data are needed for a confirmation.[17]

- *Heavy right-handed neutrinos.* Though cosmological observations place currently a stringent upper bound on neutrino masses, neutrino mixing experiments have robustly established that at least two of the three ordinary neutrinos mix and are massive. This is a very important discovery since it requires an extension of the standard model and it is the only non-cosmological evidence for new physics so far. It is then quite reasonable to think that the same extension of the standard model that can explain neutrino masses and mixing might also be able to address some of the cosmological puzzles, in particular dark matter. From this point of view the most attractive way to explain neutrino masses and mixing is introducing three right-handed neutrinos with masses M_{Ri}, much above the electroweak scale $E_{EW} \sim 100$ GeV. In this way

---

[17]It should be said, however, that a few other models of keV dark matter have been proposed to explain the anomaly.

it is possible to derive the famous see-saw relation for light neutrino masses [26], that in the simplified toy model case of just 1 active (left-handed) + 1 right-handed neutrino species reads as

$$m_\nu \sim \frac{m_{EW}^2}{M_R}. \tag{17.19}$$

The reason for the attractiveness of this relation is that one expects in grand-unified theories M_R to have values of the order of the grand-unified scale $M_{GUT} \sim (10^{15}\text{–}10^{16})\,\text{GeV}$. In this way for the light neutrino mass one obtains $m_\nu \sim 0.01\,\text{eV}$, just the scale indicated by neutrino oscillation experiments. More generally, one needs more than one right-handed neutrino species to describe all neutrino data with the seesaw formula.[18] In this case the masses can range between the electroweak scale and the grand-unified scale. These heavy right-handed neutrinos are in general highly unstable and, therefore, cannot play the role of dark matter. However, since usually in many models one expects to have three of them, at least one can be made cosmologically stable imposing a symmetry, as usual for dark matter candidates [27]. This stable right-handed neutrino will need, for its production, some feeble coupling either with the inflaton, the particle associated with the field responsible for inflation, or with some other particle that is more strongly coupled. Through this tiny coupling a small abundance of dark matter right-handed neutrinos can be generated and reproduce the observed dark matter abundance. Since the right-handed neutrinos have a huge mass, just a small abundance is needed to reproduce the observed one (see Eq. (17.8)). The same coupling is, however, typically also breaking the symmetry and the dark matter right-handed neutrinos are expected to decay. The products of these decays, in general, produce signals in indirect searches, typically very high-energy neutrinos. Neutrino telescopes place in this way a severe lower bound on the life-time of these very massive particles, of the order of $\tau \gtrsim 10^{28}\,\text{s}$, ten orders of magnitude greater than the life of the universe.

As mentioned, in 2013 the IceCube neutrino telescope, located at the South Pole, announced the detection of very high-energy (active) neutrinos with energies as high as $E_\nu \sim \text{PeV} = 10^6\,\text{GeV}$. These events are too energetic to be accounted for by known local sources like atmospheric neutrinos (and certainly not by solar neutrinos). Far astrophysical sources such as γ-ray bursts, associated with particularly energetic SN explosions, or active galactic nuclei might explain their origin but at the moment they do not seem to be able to explain the whole spectrum of observed events and, as above-mentioned, an excess at $E_\nu \sim 100\,\text{TeV}$ has been claimed. It has been proposed that some of these very high energy neutrinos are the product of the decays of very heavy

---

[18]The minimum number of right-handed neutrino species that allows to fit all current data with the seesaw formula is two, but in many models, for example in $SO(10)$ grand-unified models, one usually has three right-handed neutrino species, one for each family.

dark matter particles with mass $m_{DM} \sim$ TeV–PeV. From our discussion it should be clear that a right-handed neutrino is a natural candidate since it naturally has a mass within this range and its decays are then expected to produce neutrinos with those very high energies. At the moment it is too early to draw conclusions and a much higher amount of data is needed in order to disentangle a contribution from dark matter decays by one from conventional astrophysical sources. In particular, a dark matter contribution should manifest anisotropic features in the distribution of high-energy neutrino events, for example in the direction of the galactic centre where the density of dark matter is much higher.

- *Super massive dark matter (WIMPzillas).* Many theories, such as grand-unified theories, predict the existence of super massive particles with masses as high as the grand-unified scale, $m_X \sim (10^{15}$–$10^{16})\,\mathrm{GeV}/c^2$.[19] Since they are above the maximum allowed value of the reheat temperature, $T_{RH} \lesssim 10^{15}$ GeV, they cannot be produced thermally. However, they can be produced non-thermally at the end of the inflationary stage in different ways [28]:

  – They can be produced at reheating, when the inflaton field energy is transferred to all other particles. In this case one can show that the correct dark matter abundance can be attained even for masses that are orders of magnitude greater than the reheat temperature.

  – They can be produced by inflaton decays if the mass of the inflaton is greater than the mass of the supermassive dark matter particle.

  – Finally, they can be produced at the so-called *preheating*, when the energy of the inflaton field can create supermassive particles through non-perturbative quantum effects on a curved space leading to parametric resonant production, an effect that can be mathematically described using Bogoliubov transformation leading to a Mathieu equation for particle production. Such a preheating stage at the end of inflation would also be responsible for the production of gravitational waves during inflation and in this way it could even be tested.

Fig. 17.1, from [22], nicely summarises the variety of candidate dark matter particles that we have discussed, classifying them in terms of their mass and their coupling.[20]

This long list of candidate dark matter particles is actually only partial and it well illustrates how intense is today the investigation in finding a solution to one of the biggest and most long-standing mysteries of modern science.

---

[19]The heaviest right-handed neutrino, discussed in the previous item, might well be considered as a candidate of supermassive dark matter particle. The main point now is the considered *non-thermal* production mechanism.

[20]The case of heavy right-handed neutrinos is missing in this figure but it should basically fill the gap between the electroweak scale and the case of WIMPzillas, with very weak coupling as for the light sterile neutrinos.

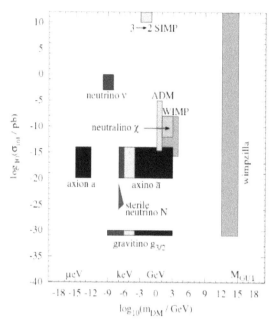

FIGURE 17.1: Summary of dark matter candidates on the plane mass versus interaction cross section (from [22]).

EXERCISES

Exercise 17.1 *Derive Eq. (17.8) for the number density of dark matter particles.*

Exercise 17.2 *Suppose that primordial black holes of mass $10^{-8} M_\odot$ make up all the dark matter in the halo of our galaxy. How far away would you expect the nearest such black hole to be? How frequently would you expect such a black hole to pass within $1 AU$ of the Sun?*

Exercise 17.3 *Imposing an instantaneous freeze-out condition on the evolution of the dark matter abundance, derive an expression for the relic abundance to be compared with Eq. (17.10).*

Exercise 17.4 *Derive Eq. (17.10) as a (freeze-out) solution for the relic abundance of the Lee–Weinberg equation (17.9).*

Exercise 17.5 *Solve numerically the Lee–Weinberg equation starting from initial thermal abundance and compare the numerical solution with the analytical solution previously found. Improve your analytical solution until you obtain an agreement within 10%.*

Exercise 17.6 *Repeat the two previous exercises but this time starting from initial vanishing abundance (freeze-in solution). Compare the thermally averaged cross* *sections required to reproduce the observed dark matter abundance in the two cases of freeze-out and freeze-in solution.*

BIBLIOGRAPHY

[1] M. G. Aartsen et al. [IceCube Collaboration], Science **342** (2013) 1242856 [arXiv:1311.5238 [astro-ph.HE]].

[2] W. Buchmuller, P. Di Bari and M. Plumacher, Annals Phys. **315** (2005) 305 [hep-ph/0401240].

[3] R. D. Peccei and H. R. Quinn, *CP Conservation in the Presence of Instantons*, Phys. Rev. Lett. **38** (1977) 1440.

[4] M. Milgrom, *A modification of the Newtonian dynamics as a possible alternative to the hidden mass hypothesis*, Astrophys. J. **270** (1983) 365.

[5] J. D. Bekenstein, *Relativistic gravitation theory for the MOND paradigm*, Phys. Rev. D **70** (2004) 083509 Erratum: [Phys. Rev. D **71** (2005) 069901] [astro-ph/0403694].

[6] B. P. Abbott et al. [LIGO Scientific and Virgo Collaborations], *Observation of gravitational waves from a binary black hole merger*, Phys. Rev. Lett. **116** (2016) 6, 061102 [arXiv:1602.03837 [gr-qc]].

[7] S. Bird, I. Cholis, J. B. Muñoz, Y. Ali-Haimoud, M. Kamionkowski, E. D. Kovetz, A. Raccanelli and A. G. Riess, *Did LIGO detect dark matter?*, Phys. Rev. Lett. **116** (2016) no.20, 201301 [arXiv:1603.00464 [astro-ph.CO]].

[8] T. D. Brandt, *Constraints on MACHO dark matter from compact stellar systems in ultra-faint dwarf galaxies*, Astrophys. J. **824** (2016) no.2, L31 [arXiv:1605.03665 [astro-ph.GA]].

[9] M. Davis, F. J. Summers and D. Schlegel, *Large scale structure in a universe with mixed hot and cold dark matter*, Nature **359** (1992) 393.

[10] P. Hut, *Limits on masses and number of neutral weakly interacting particles*, Phys. Lett. **69B** (1977) 85;

[11] B. W. Lee and S. Weinberg, *Cosmological lower bound on heavy neutrino masses*, Phys. Rev. Lett. **39** (1977) 165.

[12] M. I. Vysotsky, A. D. Dolgov and Y. B. Zeldovich, *Cosmological restriction on neutral lepton masses*, JETP Lett. **26** (1977) 188 [Pisma Zh. Eksp. Teor. Fiz. **26** (1977) 200].

[13] For a derivation of the Lee–Weinberg equation from the Boltzmann equation and the analytic calculation of the relic abundance we refer the reader to:

E. W. Kolb and M. S. Turner, *The Early universe*, Front. Phys. **69** (1990) 1, Chapter 5.

[14] G. Bertone, D. Hooper and J. Silk, *Particle dark matter: Evidence, candidates and constraints*, Phys. Rept. **405** (2005) 279 [hep-ph/0404175].

[15] J. L. Feng, *Dark matter candidates from particle physics and methods of detection*, Ann. Rev. Astron. Astrophys. **48** (2010) 495 [arXiv:1003.0904 [astro-ph.CO]].

[16] H. Goldberg, *Constraint on the photino mass from cosmology*, Phys. Rev. Lett. **50** (1983) 1419 Erratum: [Phys. Rev. Lett. **103** (2009) 099905].

[17] J. R. Ellis, J. S. Hagelin, D. V. Nanopoulos, K. A. Olive and M. Srednicki, *Supersymmetric relics from the Big Bang*, Nucl. Phys. B **238** (1984) 453.

[18] L. Roszkowski, E. M. Sessolo and S. Trojanowski, *WIMP dark matter candidates and searches - current issues and future prospects*, Rep. Prog. Phys., (2018), doi.org/10.1088/1361-6633/aab913 [arXiv:1707.06277 [hep-ph]].

[19] M. G. Aartsen et al. [IceCube Collaboration], *First observation of PeV-energy neutrinos with IceCube*, Phys. Rev. Lett. **111** (2013) 021103 [arXiv:1304.5356 [astro-ph.HE]].

[20] M. Chianese, G. Miele and S. Morisi, *Interpreting IceCube 6-year HESE data as an evidence for hundred TeV decaying Dark Matter*, Phys. Lett. B **773** (2017) 591 [arXiv:1707.05241 [hep-ph]].

[21] K. Griest and D. Seckel, *Three exceptions in the calculation of relic abundances*, Phys. Rev. D **43** (1991) 3191.

[22] H. Baer, K. Y. Choi, J. E. Kim and L. Roszkowski, *Dark matter production in the early Universe: beyond the thermal WIMP paradigm*, Phys. Rept. **555** (2015) 1 [arXiv:1407.0017 [hep-ph]].

[23] L. J. Hall, K. Jedamzik, J. March-Russell and S. M. West, *Freeze-In Production of FIMP Dark Matter*, JHEP **1003** (2010) 080 [arXiv:0911.1120 [hep-ph]].

[24] M. Y. Khlopov and A. D. Linde, Phys. Lett. B **138** (1984) 265; J. R. Ellis, J. E. Kim and D. V. Nanopoulos, Phys. Lett. B **145** (1984) 181; K. Kohri, T. Moroi and A. Yotsuyanagi, Phys. Rev. D **73** (2006) 123511; M. Kawasaki, K. Kohri, T. Moroi and A. Yotsuyanagi, Phys. Rev. D **78** (2008) 065011 [arXiv:0804.3745 [hep-ph]].

[25] For a review on the 3.5 keV line and its interpretation in terms of sterile neutrino dark matter see: K. N. Abazajian, *Sterile neutrinos in cosmology*, arXiv:1705.01837 [hep-ph].

[26] P. Minkowski, Phys. Lett. B **67** (1977) 421; T. Yanagida, in Proceedings of the Workshop on Unified Theory and Baryon Number of the Universe, eds. O. Sawada and A. Sugamoto (KEK, 1979) p.95; P. Ramond, Invited talk given at Conference: C79-02-25 (Feb 1979) p.265-280, CALT-68-709, hep-ph/9809459; M. Gell-Mann, P. Ramond and R. Slansky, in Supergravity, eds. P. van Niewwenhuizen and D. Freedman (North Holland, Amsterdam, 1979) Conf.Proc. C790927 p.315, PRINT-80-0576; R. Barbieri, D. V. Nanopoulos, G. Morchio and F. Strocchi, Phys. Lett. B **90** (1980) 91; R. N. Mohapatra and G. Senjanovic, Phys. Rev. Lett. **44** (1980) 912.

[27] A. Anisimov and P. Di Bari, *Cold dark matter from heavy right-handed neutrino mixing*, Phys. Rev. D **80** (2009) 073017 [arXiv:0812.5085 [hep-ph]].

[28] D. J. H. Chung, E. W. Kolb and A. Riotto, *Superheavy dark matter*, Phys. Rev. D **59** (1999) 023501 [hep-ph/9802238]; V. Kuzmin and I. Tkachev, *Ultrahigh-energy cosmic rays, superheavy long living particles, and matter creation after inflation*, JETP Lett. **68** (1998) 271 [Pisma Zh. Eksp. Teor. Fiz. **68** (1998) 255] [hep-ph/9802304].

CHAPTER 18

Ad libitum?

In our journey through modern cosmology, we could see how long-standing issues have finally found a solution. We have a consistent cosmological model that explains all current cosmological observations, from the expansion of the universe to the properties of the CMB, from the primordial nuclear abundances to the distribution of galaxies. Our observable universe seems from many points of view a much less mysterious place than a hundred years ago, at the time of the Great Debate, when it was not clear whether our galaxy could encompass the entire universe. Yet, we ended our journey with new puzzles that, despite formidable efforts, do not seem to have easy solutions. Cosmology is today much more tightly connected to fundamental physics than in the past and in fact it seems to provide a powerful phenomenological tool to circumvent current difficulties to access higher and higher energies in laboratory experiments. From this point of view cosmology looks like a field that, for a long time, will manifest a huge potential for scientific discoveries. The dark matter of the universe seems a much more difficult puzzle to be solved than was thought a few decades ago, when there was a positivistic expectation that a new powerful collider would have disclosed the mysteries of the universe. Negative results in dark matter searches seem to confirm that our idea of what should be *natural* does not seem to apply to Nature in a straightforward way, but is much more subtle and difficult to read than we thought.

Today it seems actually that the universe gives us the tools to explore new territories beyond the standard model. In particular, the discovery of gravitational waves might prove to be a revolutionary new way to explore the universe and finally give us access to stages in the history of the universe prior to Big Bang nucleosynthesis.

Neutrino physics seems to suggest interesting solutions to the cosmological puzzles that combine well with cosmological observations so far. The mystery of the matter-antimatter asymmetry of the universe might indeed be related to neutrino properties and future neutrino experiments might shed light on it. Vice versa cosmological observations might finally measure the absolute neutrino mass scale, playing a very important role in models of leptogenesis.

Therefore, it seems that the music tune of new cosmological discoveries will still play for long time, apparently able to surprise us *ad libitum* with its exciting twists. Certainly at the moment, with so many unanswered puzzles, a *gran finale* seems still to be far to come. For this reason, cosmology still represents today, a hundred years since its birth, one of the most exciting scientific adventures.

CHAPTER 19

Summary of numerical values of constants and parameters

UNITS

Astronomical Unit :	$1\,\mathrm{AU} = 1.496 \times 10^{11}\,\mathrm{m} \simeq 150 \times 10^{6}\,\mathrm{km}$
Light-year:	$1\,\mathrm{ly} = 0.9467 \times 10^{16}\,\mathrm{m} \simeq 10,000 \times 10^{9}\,\mathrm{km}$
Parsec:	$1\,\mathrm{pc} = 3.0856 \times 10^{16}\,\mathrm{m} = 3.2615\,\mathrm{ly} = 30,856 \times 10^{9}\,\mathrm{km}$
Megaparsec:	$1\,\mathrm{Mpc} \equiv 10^{6}\,\mathrm{pc} \simeq 3.0856 \times 10^{22}\,\mathrm{m}$
Sidereal Gigayear:	$1\,\mathrm{Gyr} = 3.16 \times 10^{16}\,\mathrm{s}$
Electronvolt:	$1\,\mathrm{eV} = 1.6022 \times 10^{-19}\,\mathrm{J}$
Mass :	$1\,\mathrm{eV}/c^{2} = 1.7827 \times 10^{-36}\,\mathrm{kg}$
Temperature-Energy:	$1\,\mathrm{K} = 0.8617 \times 10^{-4}\,\mathrm{eV}$
Cross section:	$1\,\mathrm{barn} = 10^{-28}\,\mathrm{m}^{2}$

FUNDAMENTAL CONSTANTS

speed of light:	$c = 299792458\,\mathrm{m\,s^{-1}}$
Planck constant $\times\ c$:	$\hbar c = 0.19733\,\mathrm{GeV} \times 10^{-15}\,\mathrm{m}$
Planck mass:	$M_P = 1.2211 \times 10^{19}\,\mathrm{GeV}/c^{2}$
gravitational constant:	$G = \hbar c / M_P^{2} = 6.7065 \times 10^{-38}\,(\mathrm{GeV}/c^{2})^{-2}\,\hbar c$
	$= 6.6741 \times 10^{-11}\,\mathrm{m^{-3}\,kg^{-1}\,s^{-2}}$
	$\simeq 4.3 \times 10^{-9}\,\mathrm{km^{2}\,s^{-2}\,Mpc}\,M_\odot^{-1}$
fine-structure constant:	$\alpha = e^{2}/(4\pi\,\hbar c) = 1/137.036$
Fermi constant:	$G_F/(\hbar c)^{3} = 1.1663787 \times 10^{-5}\,\mathrm{GeV}^{-2}$
Thomson cross section:	$\sigma_e = 6.6524 \times 10^{-29}\,\mathrm{m}^{2}$
electron mass:	$m_e = 0.511\,\mathrm{MeV}/c^{2}$
proton mass:	$m_p = 938.3\,\mathrm{MeV}/c^{2}$
neutron mass:	$m_n = 939.6\,\mathrm{MeV}/c^{2}$

COSMOLOGICAL PARAMETERS

The values of parameters from CMB anisotropies (e.g., energy-matter budget parameters) with their errors are those obtained by the *Planck* collaboration ('TT+lowP+lensing') within the ΛCDM model for $\Omega_0 = 1$ or, in the case of Ω_0, within an extended ($\Omega_0 + \Lambda$CDM) model (see Tables 12.1 and 12.2 respectively for a reference). For H_0 we also give the value found by the HST collaboration.

$$\text{Hubble constant:} \quad H_0 = (67.8 \pm 0.9)\,\text{km s}^{-1}\,\text{Mpc}^{-1}$$

$$\text{Hubble time at present:} \quad t_{H,0} \equiv H_0^{-1} = (14.4 \pm 0.2)\,\text{Gyr}$$

$$\text{Hubble radius at present:} \quad R_{H,0} \equiv c\,H_0^{-1} = (4.42 \pm 0.06)\,\text{Gpc}$$

$$\text{Hubble constant (HST):} \quad H_0 = (73.24 \pm 1.74)\,\text{km s}^{-1}\,\text{Mpc}^{-1}$$

$$\text{Age of the universe:} \quad t_0 = (13.8 \pm 0.04)\,\text{Gyr}$$

$$\text{Temperature of CMB:} \quad T_0 = (2.7255 \pm 0.0006)\,K$$
$$= (2.3486 \pm 0.0005) \times 10^{-4}\,\text{eV}$$

$$\text{number density of relic photons:} \quad n_{\gamma,0} \simeq 410.7\,\text{cm}^{-3}$$

$$\text{energy density of relic photons:} \quad \varepsilon_{\gamma,0} \simeq 0.2602\,\text{MeV m}^{-3}$$

$$\text{critical energy density at present:} \quad \varepsilon_{c,0} \simeq 10.54\,h^2\,\text{GeV m}^{-3}$$

$$\text{energy density parameter at present:} \quad \Omega_0 = 1.005^{+0.016}_{-0.017}$$

$$\text{baryons:} \quad \Omega_{B,0}h^2 = 0.02226 \pm 0.00023$$
$$\Omega_{B,0} = 0.048 \pm 0.002$$

$$\text{baryon-to-photon ratio:} \quad \eta_{B,0} = (6.09 \pm 0.06) \times 10^{-10}$$

$$\text{matter:} \quad \Omega_{M,0}h^2 = 0.1415 \pm 0.0019$$
$$\Omega_{M,0} = 0.308 \pm 0.012$$

$$\text{(cold) dark matter:} \quad \Omega_{DM,0}h^2 = 0.1186 \pm 0.0020$$
$$\Omega_{DM,0} = 0.258 \pm 0.011$$

$$\text{cosmological constant:} \quad \Omega_{\Lambda,0} = 0.692 \pm 0.012$$

$$\text{radiation} \quad \Omega_{R,0} \simeq 0.915 \times 10^{-4}$$

$$\text{matter-}\Lambda\text{ equality:} \quad t_{eq}^{M\Lambda} \simeq (10.2 \pm 0.2)\,\text{Gyr}\,,\ a_{eq}^{M\Lambda} \simeq 0.75$$

$$\text{matter-radiation decoupling:} \quad t_{dec} \simeq 371,500\,\text{yr}\,,\ z_{dec} \simeq 1090,$$
$$T_{dec} \simeq 0.256\,\text{eV}$$

$$\text{matter-radiation equality:} \quad t_{eq} \simeq 52,000\,\text{yr}\,,\ z_{eq} = 3365 \pm 44\,,$$
$$T_{eq} \simeq 0.80\,\text{eV}$$

$$\text{nucleosynthesis:} \quad t_{nuc} \simeq 310\,\text{s}\,,\ T_{nuc} \simeq 0.065\,\text{MeV}$$

$$n/p\text{ freeze-out:} \quad t_f \simeq 1.0\,\text{s}\,,\ T_f \simeq 0.85\,\text{MeV}$$

ASTROPHYSICAL PARAMETERS

$$\text{solar mass:} \quad M_\odot = 1.989 \times 10^{30}\,\text{kg}$$

$$\text{solar radius:} \quad R_\odot = 695,980\,\text{km}$$

$$\text{solar luminosity:} \quad L_\odot = 3.828 \times 10^{26}\,\text{W}$$

Index

Milton Keynes UK
Ingram Content Group UK Ltd.
UKHW051950071024
449327UK00026B/2250